Cabin Crew Conference 2018

Safety Through Efficient Training

Held at APATS 2018

Singapore
28 - 29 August 2018

ISBN: 978-1-7138-0595-3

Printed from e-media with permission by:

Curran Associates, Inc.
57 Morehouse Lane
Red Hook, NY 12571

Some format issues inherent in the e-media version may also appear in this print version.

Printed by Curran Associates, Inc. (2020)

For permission requests, please contact Halldale Media Inc.
at the address below.

Halldale Media Inc.
735 Primera Blvd.
Suite 220
Lake Mary, Florida 32746
USA

Phone: (407) 322-5605
Fax: (407) 322-5604

www.halldale.com

Additional copies of this publication are available from:

Curran Associates, Inc.
57 Morehouse Lane
Red Hook, NY 12571 USA
Phone: 845-758-0400
Fax: 845-758-2633
Email: curran@proceedings.com
Web: www.proceedings.com

TABLE OF CONTENTS

IATA
APATS 2018 - Singapore

Blair Cowles
Regional Director – Safety and Flight Operations
Asia Pacific

Today

- Blair Cowles – Regional Director, Safety & Flight Operations Asia Pacific
- What is IATA
- Economic overview
- IATA initiatives in the training domain

IATA Overview

- Global trade association for the world's airlines
- 287 passenger and cargo carriers in 120 countries
- 84% of global air traffic
- 54 offices in 53 countries with 1,500 employees
- Founded in Havana, Cuba in April 1945
- Head Office: Montreal, Canada
  Executive Office: Geneva, Switzerland

IATA Overview

- Mission: To represent, lead and serve the airline industry
- Leadership:
  - Chair of IATA Board of Governors – QR CEO Akbar Al Baker
  - IATA Director General and CEO – Alexandre de Juniac

IATA Overview

BOARD OF GOVERNORS	31	Member airline CEOs
INDUSTRY COMMITTEES	6	Cargo, Environment, Financial, Industry Affairs, Legal & Operations
WORKING GROUPS	As Needed	Committee sub-groups established to focus on specific issues
TRAFFIC CONFERENCES	All	All members Passenger & Cargo

IATA Overview

HQ
Montreal/Geneva

Regions	Organizational Divisions
North Asia (Beijing)	Member & External Relations (MER)
ASPAC (Singapore)	Financial & Distribution Services (FDS)
AME (Amman)	Airport, Passenger, Cargo & Security (APCS)
Europe (Madrid)	Safety & Flight Operations (SFO)
Americas (Miami)	Marketing & Commercial Services (MACS)

IATA Asia Pacific

IATA Asia Pacific

Every day in 2017…

11.2 MILLION PASSENGERS

114,500 FLIGHTS

$16.6 BILLION WORTH OF GOODS CARRIED

IATA APATS - August 2018

Industry profitability still high and now stabilizing

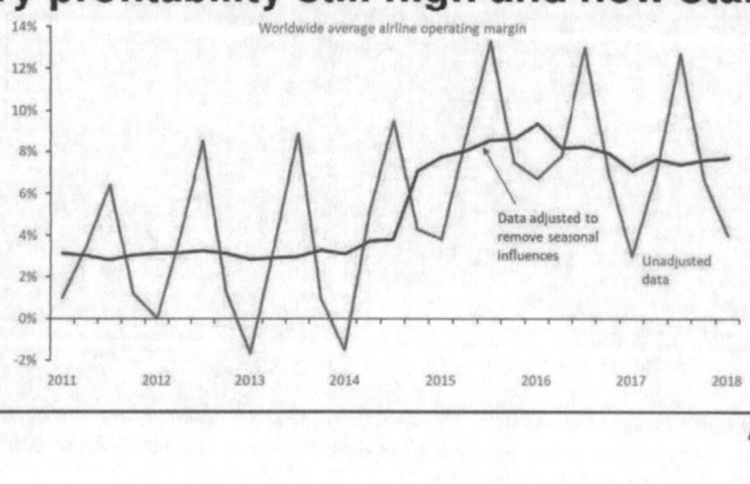

IATA APATS - August 2018

But the industry now faces late-cycle cost pressures

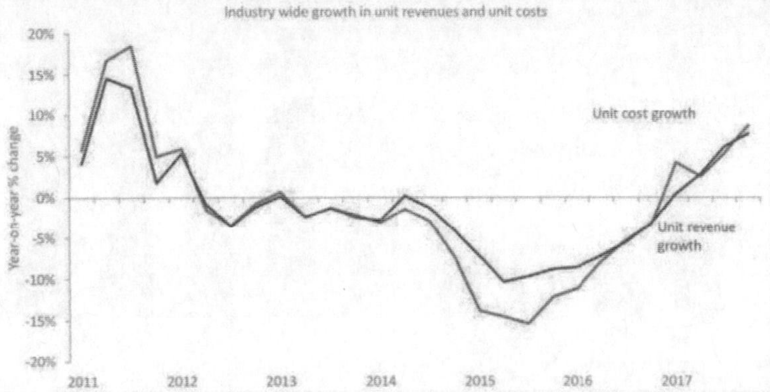

IATA APATS - August 2018

Regional profitability remains very uneven

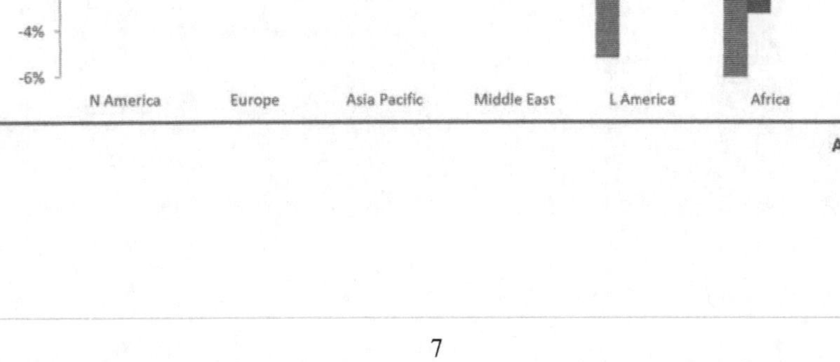

IATA APATS - August 2018

Driven largely by N.America (again) plus Europe & Asia

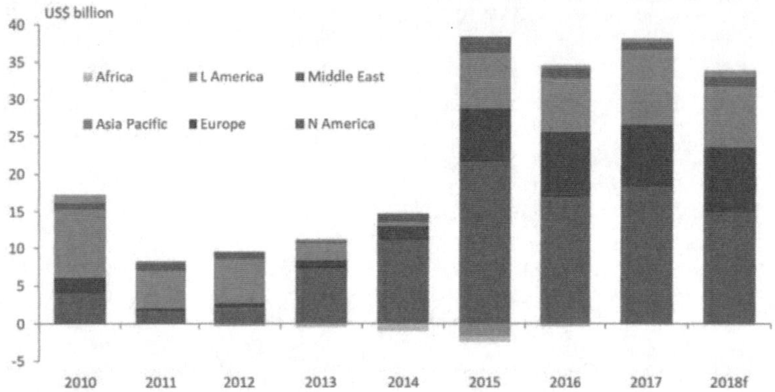

Regional (net) profitability – Asia Pacific

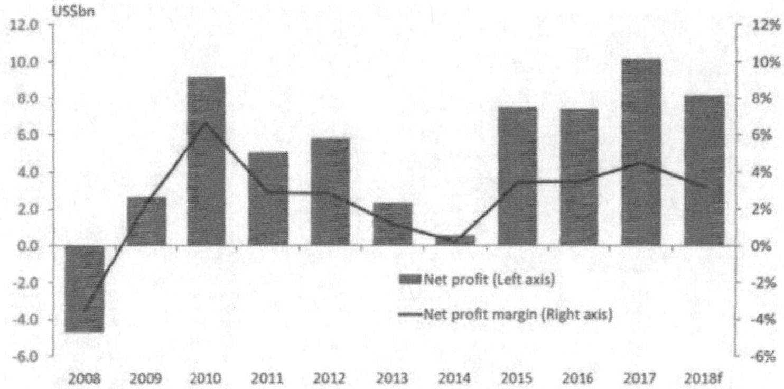

Per passenger (net) profit shows wide regional variation

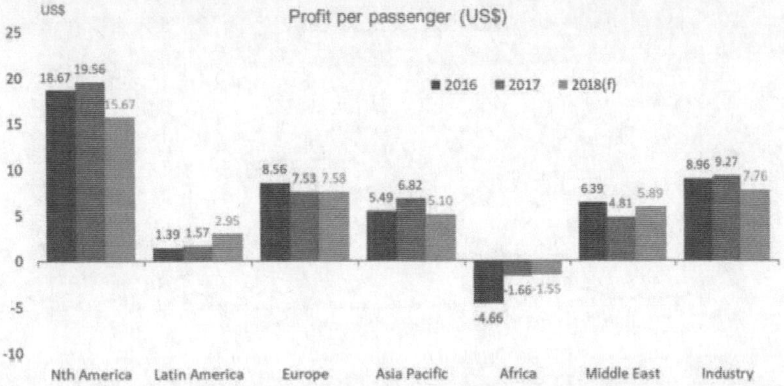

Record # of city-pairs = more choice + economic benefit

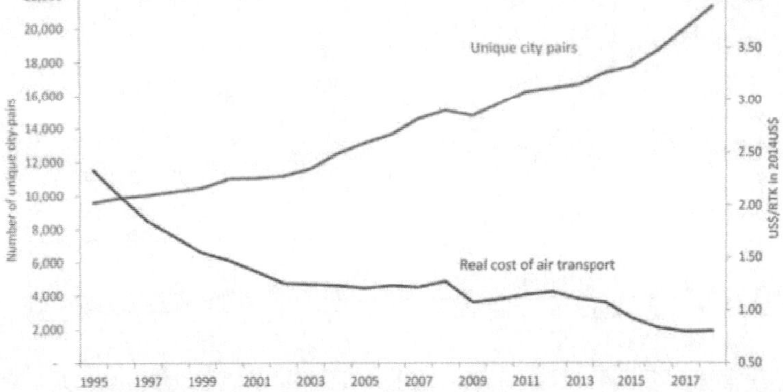

IATA

Increasing accessibility of air travel

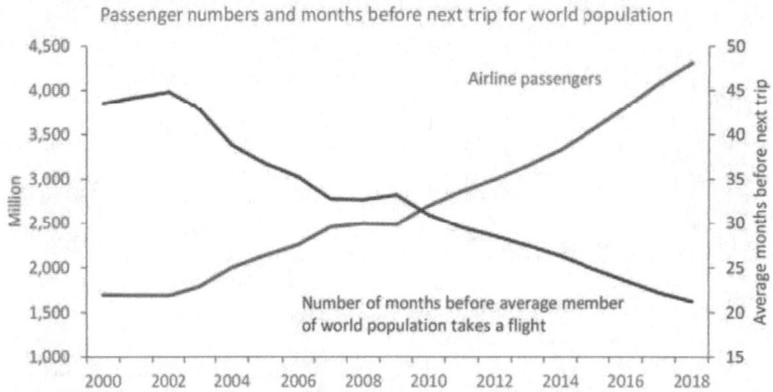

IATA

Average price of air travel (real terms)

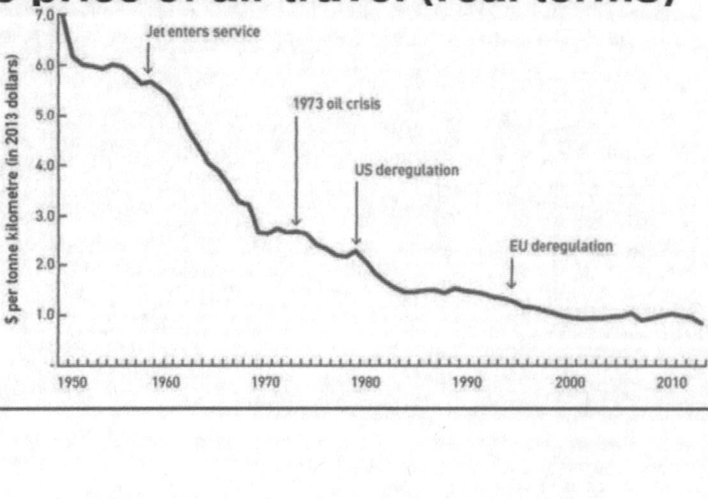

Boosting trade and tourism

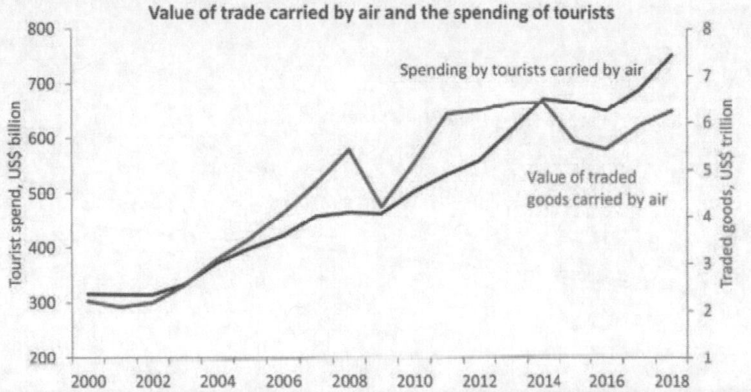

IATA APATS - August 2018

2.9 million airline jobs in 2018

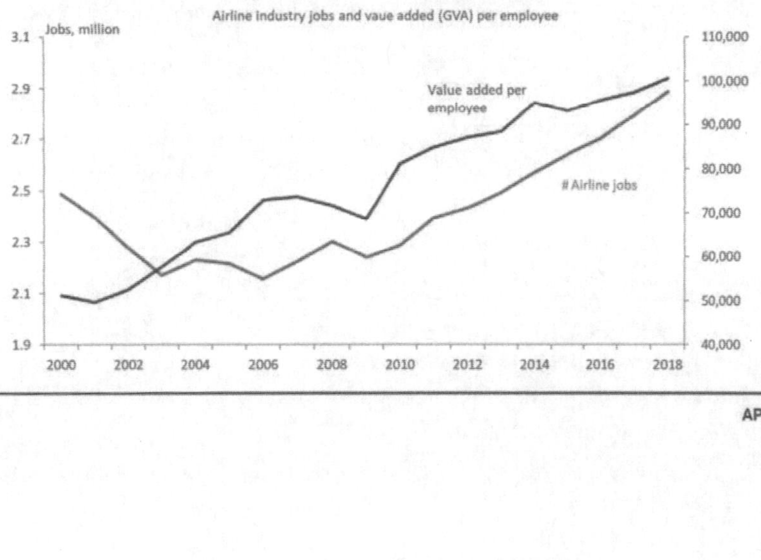

IATA APATS - August 2018

Demand for air travel to double over the next 20 years

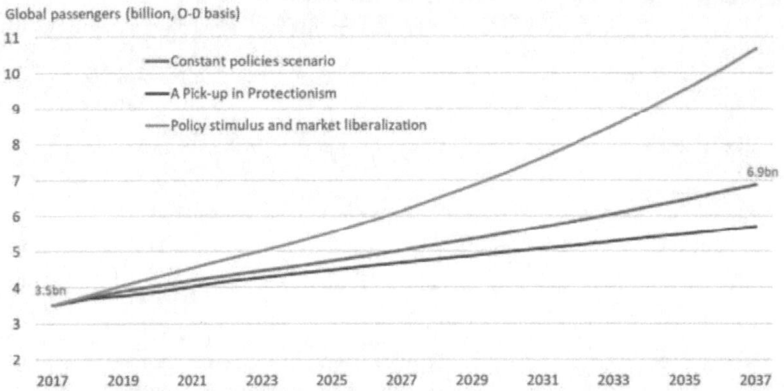

IATA

APATS - August 2018

The value of aviation in Asia-Pacific

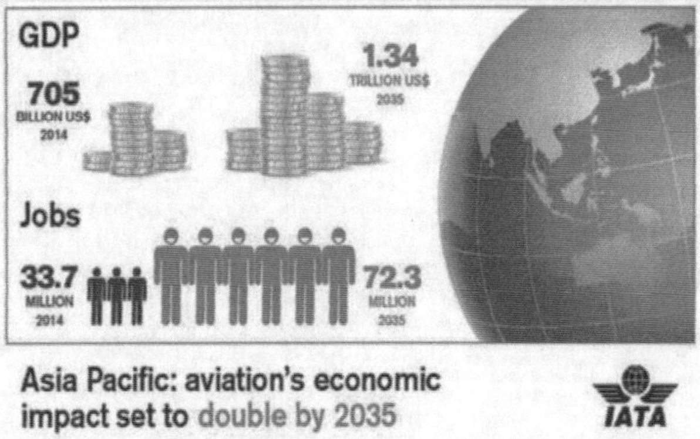

IATA

APATS - August 2018

20-year passenger growth outlook (2017-37)

IATA

APATS - August 2018

Top 10 air passenger markets 2017-37

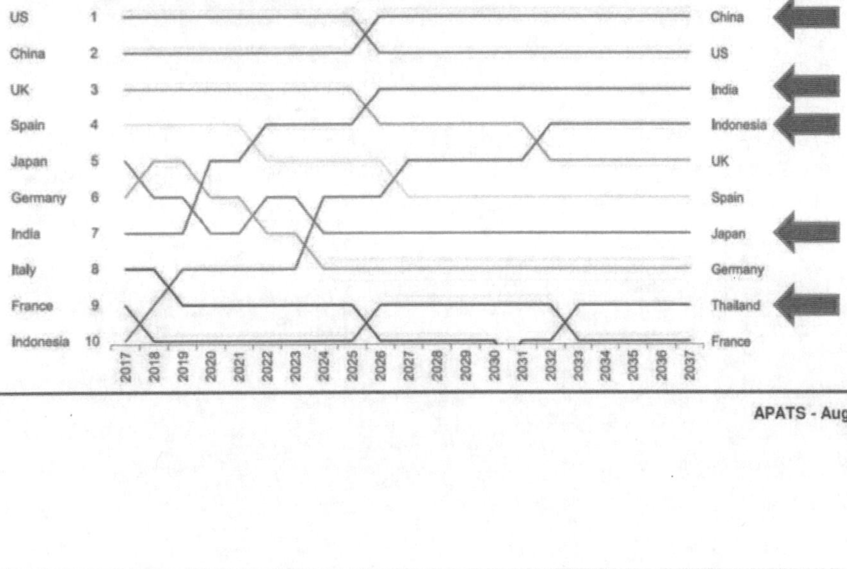

IATA

APATS - August 2018

Rise of Asian domestic markets

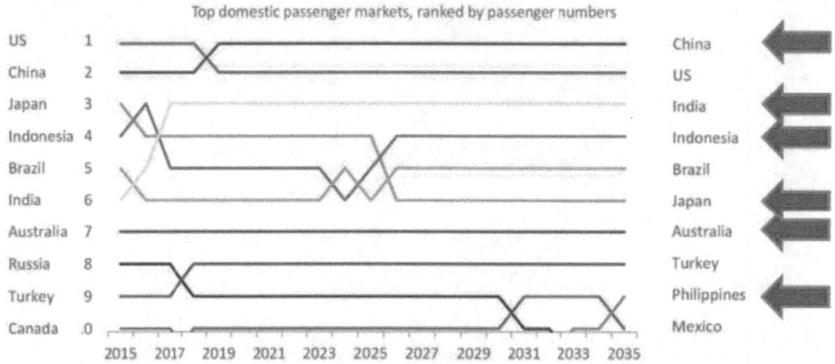

Changes in market share over time

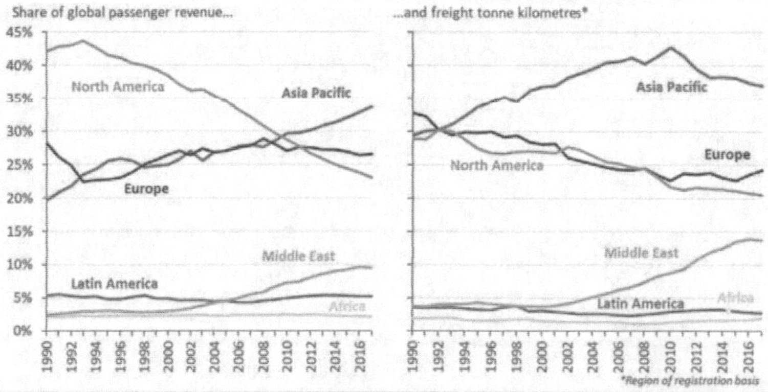

IATA in the Training Domain

IATA Governance Structure

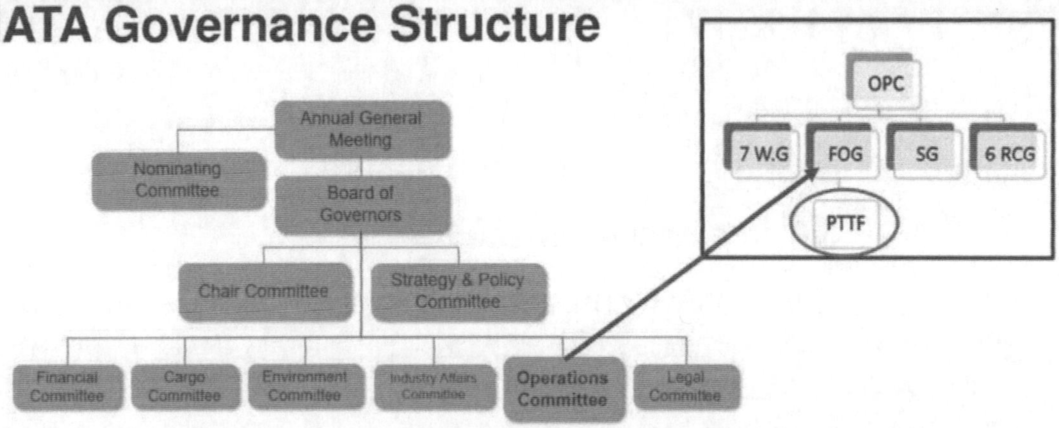

IATA APATS - August 2018

2007-2013 ITQI Total System Approach

- From the selection criteria to training process:
  - Selection Criteria (Pilot Aptitude Testing)
  - Multi-Crew Pilot License (MPL)
  - Evidence-Based Training (EBT)
  - Instructor Qualification (IQ)
  - Flight Simulation Training Devices (FSTD)
- Engineering & Maintenance (competency-based training and qualification requirements)

PTTF 2014-2017

- UPRT Guidance material
- Flight Crew Monitoring Guidance material
- TEM/CRM/Competencies model
- Instructor Evaluator Competency-Based Guidance Material

http://www.iata.org/whatwedo/ops-infra/training-licensing/Pages/index.aspx

Pilot Shortage Forecast

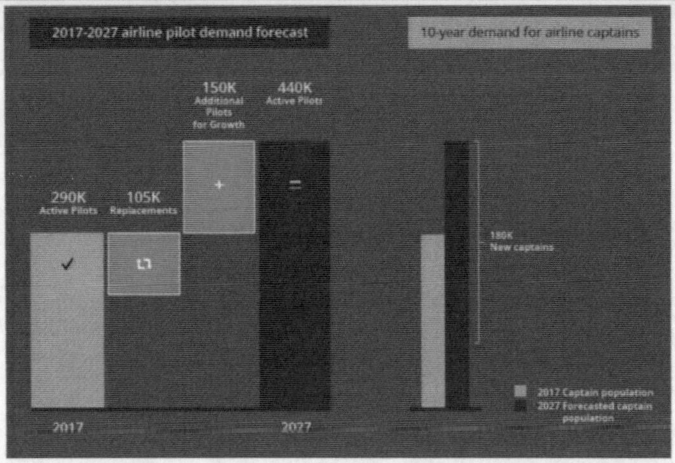

Source: CAE forecast

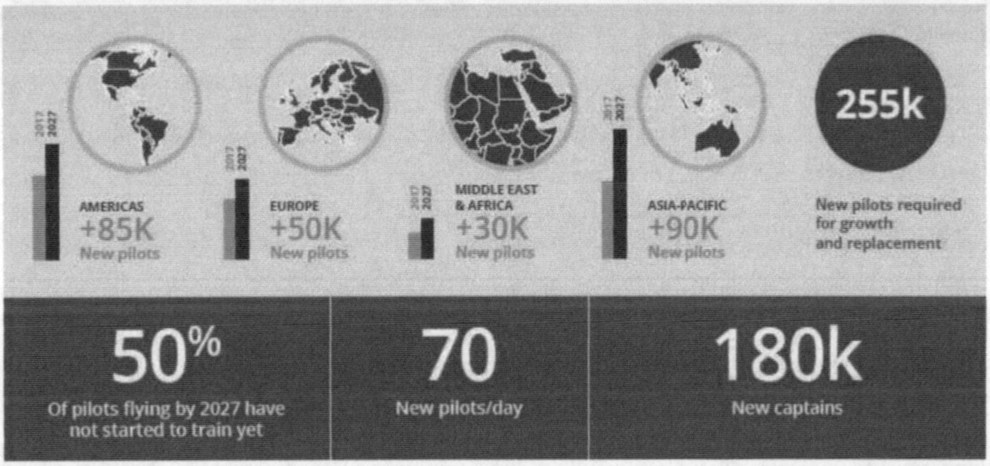

Source CAE forecast

IATA Action Plan
Phase 1: collect information

- Training Capacity Data
- Regulatory Requirements
- Career Path
- Pilot Aptitude Testing
- Approved Training Organization and Operator Partnership

IATA APATS - August 2018

IATA Action Plan
Phase 2: to mitigate

- Advocate
- Attract
- Harmonized Pilot Aptitude Testing
- CBTA implementation
- Improve ATO/Operator Partnership

IATA APATS - August 2018

From this situation

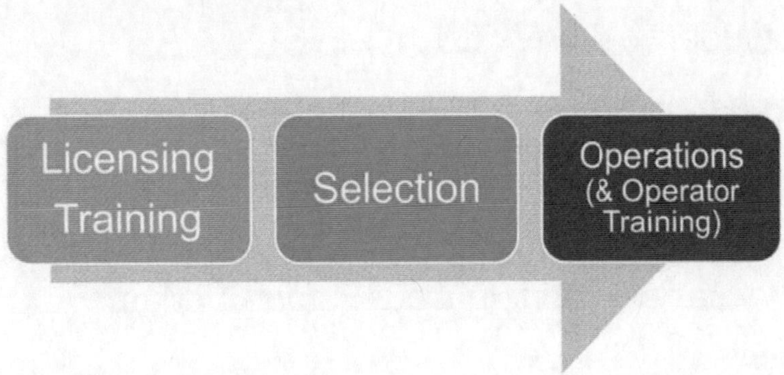

To this situation

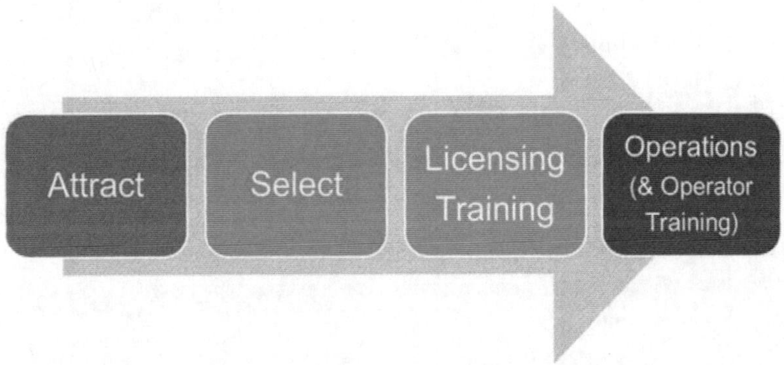

IATA

Training and Licensing Dept. 2018 priorities

- Pilot Aptitude Testing (REV 3)
- Command Course
- Competency-Based Training and Assessment (CBTA) implementation guidance
- Evidence-Based Training Data Report update
- UPRT (REV 2)

IATA APATS - August 2018

IATA

Other initiatives

- Support EBT implementation
- Prepare CBTA implementation
- Link ATO-AOC (ATO Training Performance Audit)
- New training media / methods credits

IATA APATS - August 2018

Thank You

To represent, lead and serve the airline industry

15TH ASIA PACIFIC AIRLINE TRAINING SYMPOSIUM
28 AUG 2018
Captain C E Quay
SENIOR VICE PRESIDENT FLIGHT OPERATIONS
SINGAPORE AIRLINES

AIRBUS
EMBRY-RIDDLE
Aeronautical University

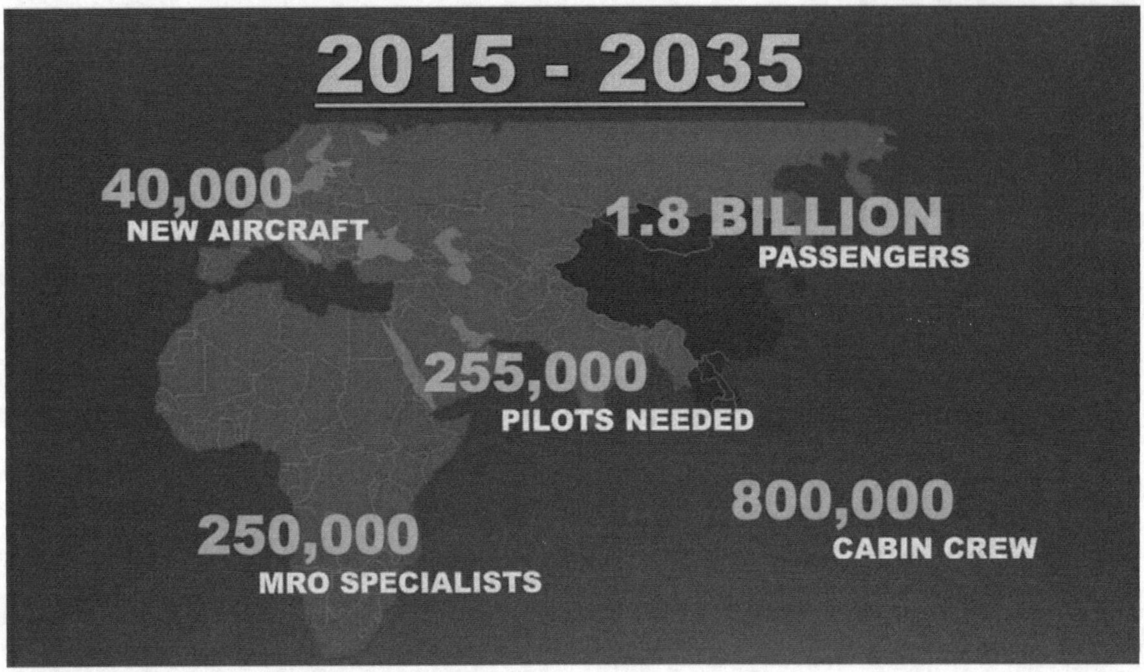
2015 - 2035
40,000
NEW AIRCRAFT
1.8 BILLION
PASSENGERS
255,000
PILOTS NEEDED
800,000
CABIN CREW
250,000
MRO SPECIALISTS

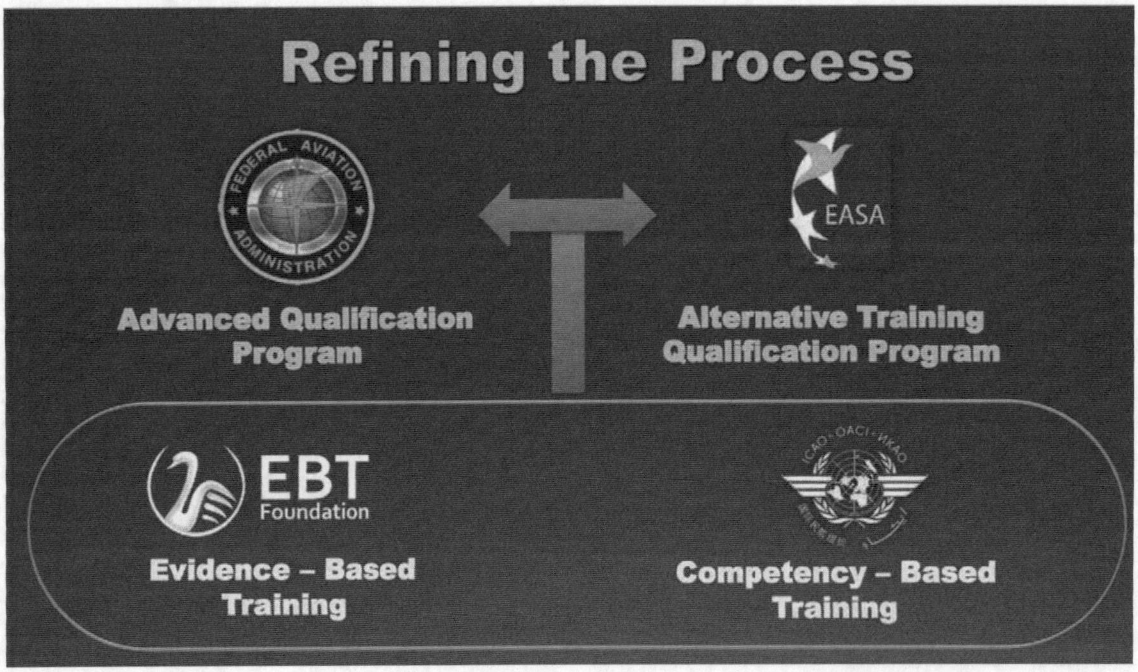
Refining the Process
FEDERAL AVIATION ADMINISTRATION
EASA
Advanced Qualification Program
Alternative Training Qualification Program
EBT Foundation
ICAO · OACI · ИКАО
Evidence – Based Training
Competency – Based Training

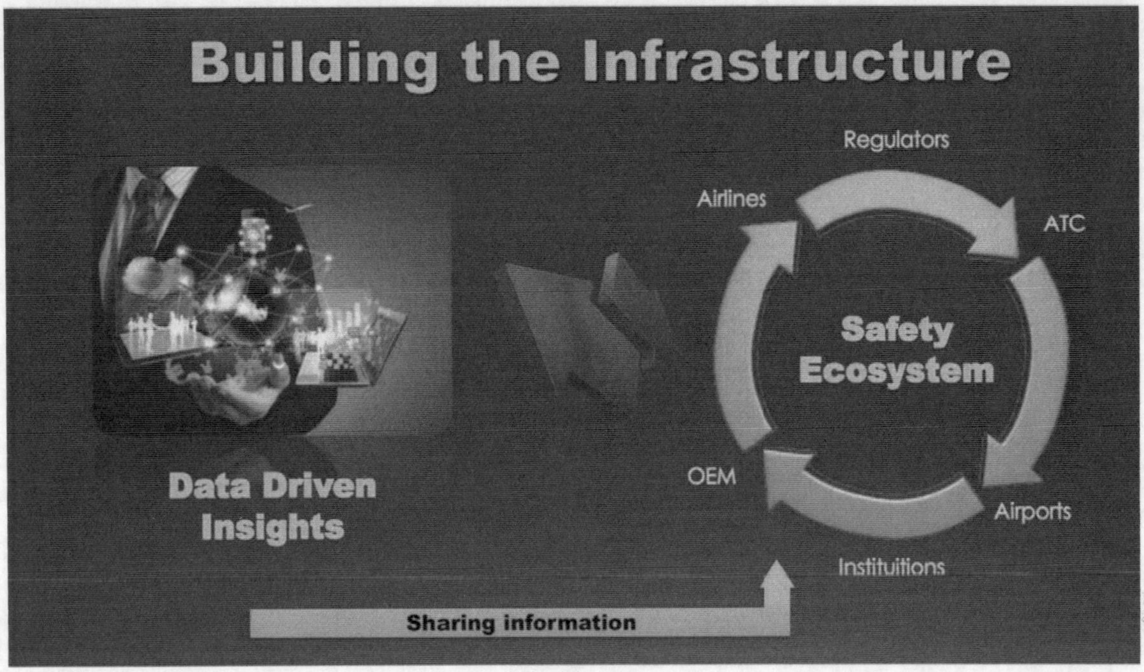
Building the Infrastructure
Regulators
Airlines
ATC
Safety Ecosystem
Data Driven Insights
OEM
Airports
Instituitions
Sharing information

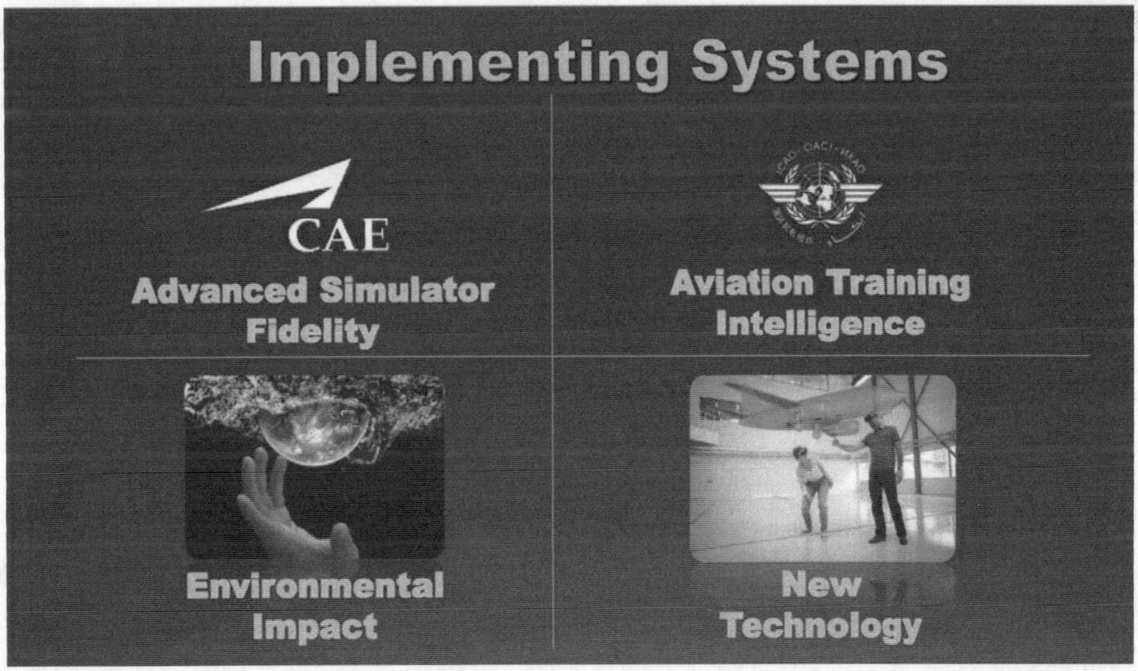
Implementing Systems
CAE
Advanced Simulator Fidelity
Aviation Training Intelligence
Environmental Impact
New Technology

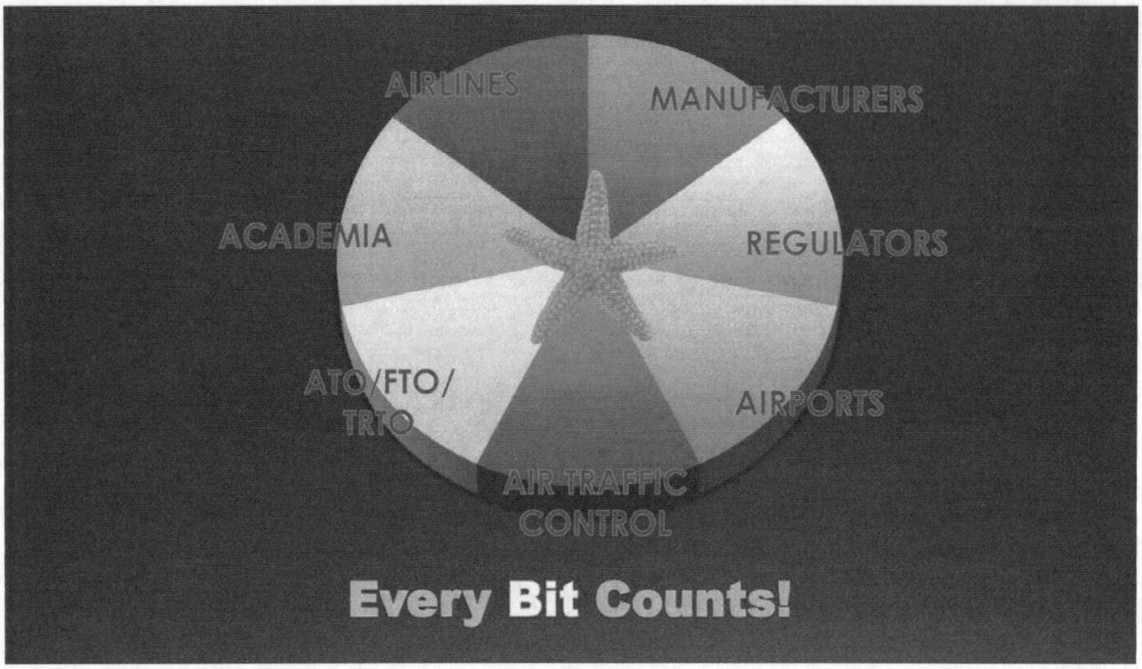
AIRLINES
MANUFACTURERS
ACADEMIA
REGULATORS
ATO/FTO/
TRTO
AIRPORTS
AIR TRAFFIC
CONTROL
Every Bit Counts!

BOEING

Global Services

2018 APAC Pilot & Technician Outlook

Patrick Curtin | Senior Manager, Singapore Campus Operations

August 28, 2018

Overview

Boeing Global Services

A dedicated services business focused on the needs of global defense, space, and commercial customers

Worldwide presence

300+ locations

70+ countries

40+ US states

Headquartered in the Dallas, TX area

Approximately 20,000 employees

Government and commercial business areas

Robust service capabilities

Digital Aviation & Analytics

Engineering, Modifications & Maintenance

Training & Professional Services

Supply Chain

Efficient performance

Faster customer response

Operational agility

Competitive cost structures

Airplane market trends

Balanced growth

Productivity and innovation

Replacement

Airlines will need 42,700 new airplanes valued at $6.3 trillion

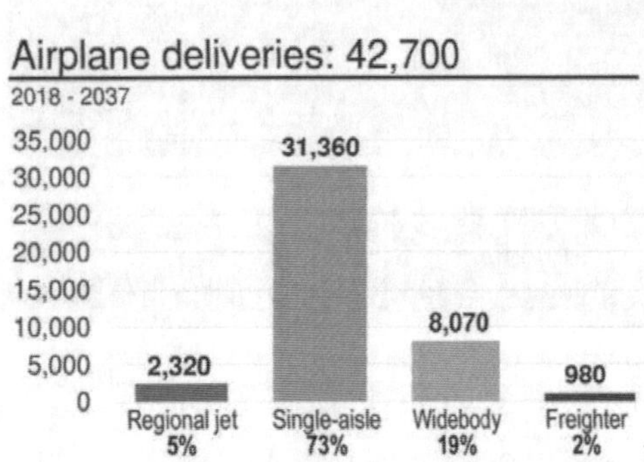

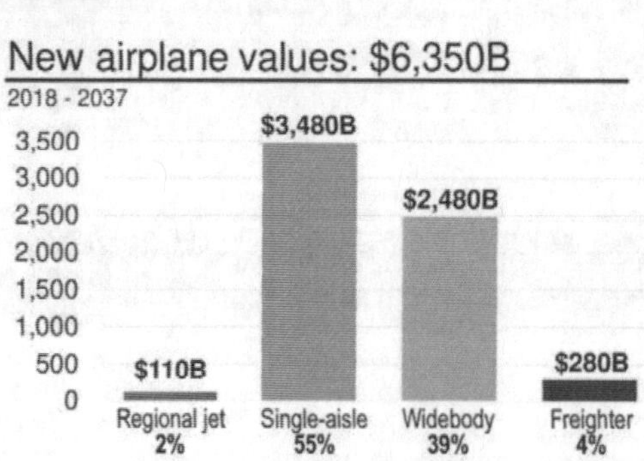

NOTE: Market share may not equal 100% due to rounding

Market for new airplanes to become even more geographically balanced

New airplane deliveries by region

2018–2037

Region	Airplanes
Asia	16,930
North America	8,800
Europe	8,490
Middle East	2,990
Latin America	3,040
Russia/CA	1,290
Africa	1,190
World Total	42,730

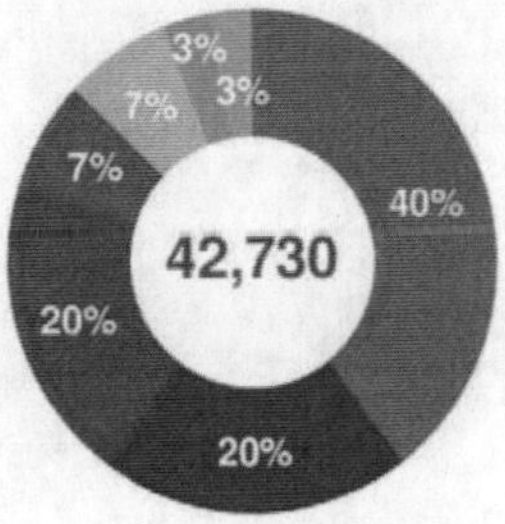

Market value by region

2018–2037

Region	$B
Asia	2,670
North America	1,100
Europe	1,240
Middle East	660
Latin America	360
Russia/CA	150
Africa	170
World Total	$6,350B

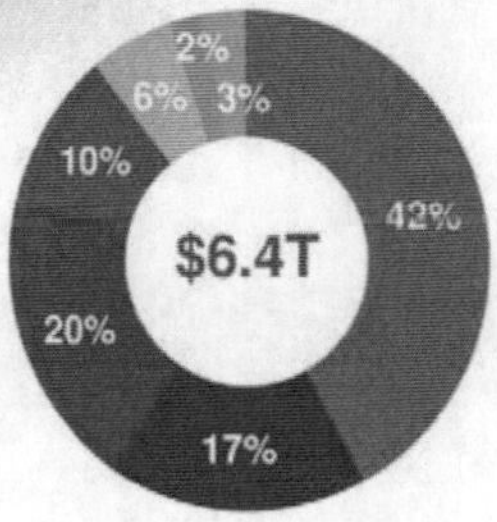

Pilot and Technician Outlook methodology

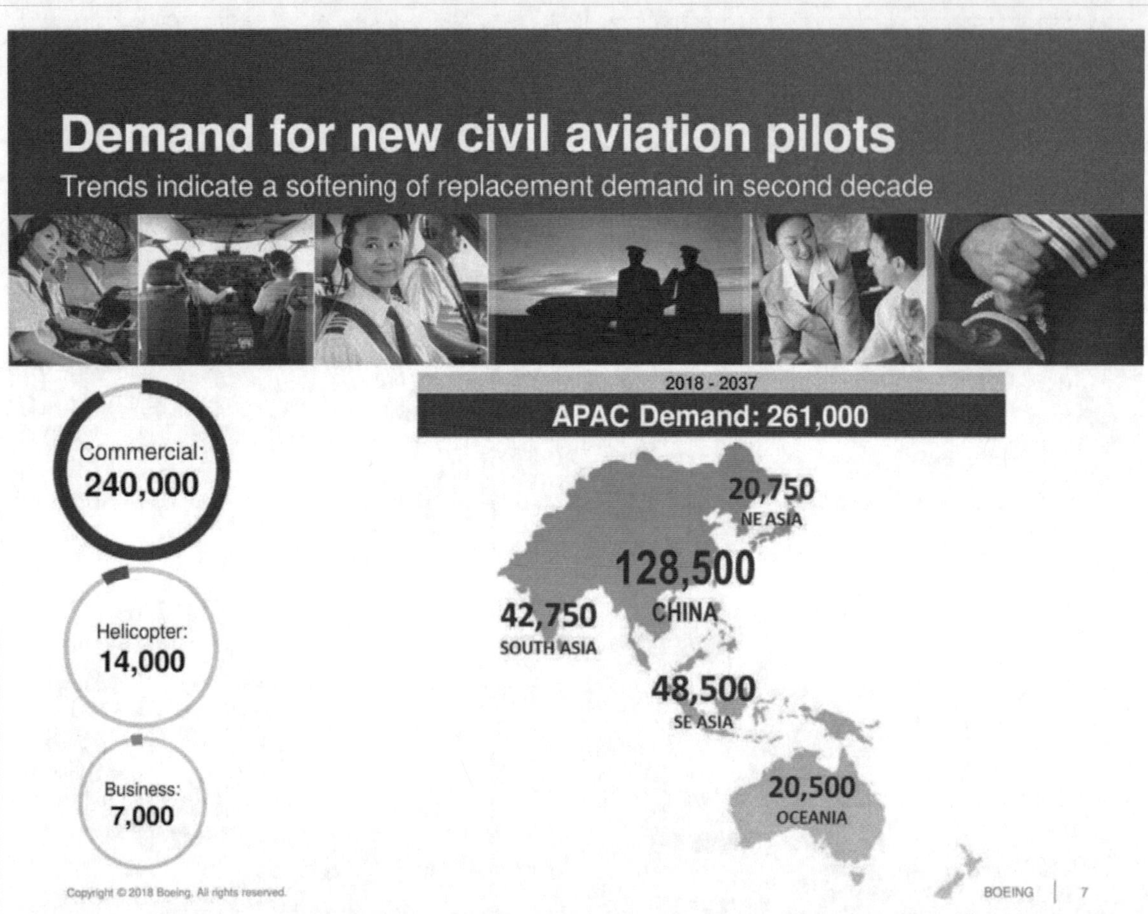
Demand for new civil aviation pilots
Trends indicate a softening of replacement demand in second decade
Commercial: 240,000
Helicopter: 14,000
Business: 7,000
2018 - 2037
APAC Demand: 261,000
20,750
NE ASIA
128,500
CHINA
42,750
SOUTH ASIA
48,500
SE ASIA
20,500
OCEANIA
Copyright © 2018 Boeing. All rights reserved.
BOEING
7

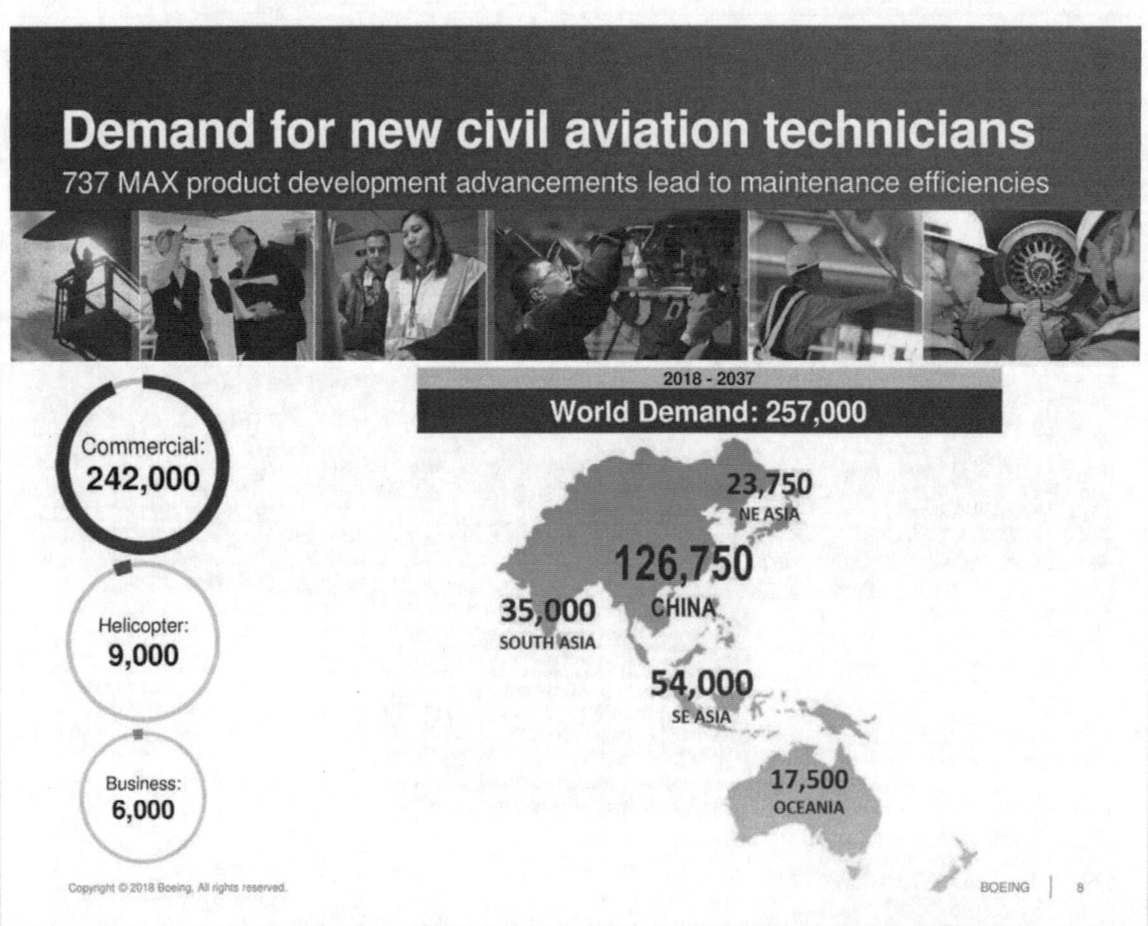
Demand for new civil aviation technicians
737 MAX product development advancements lead to maintenance efficiencies
Commercial: 242,000
Helicopter: 9,000
Business: 6,000
2018 - 2037
World Demand: 257,000
23,750
NE ASIA
126,750
CHINA
35,000
SOUTH ASIA
54,000
SE ASIA
17,500
OCEANIA
Copyright © 2018 Boeing. All rights reserved.
BOEING
8

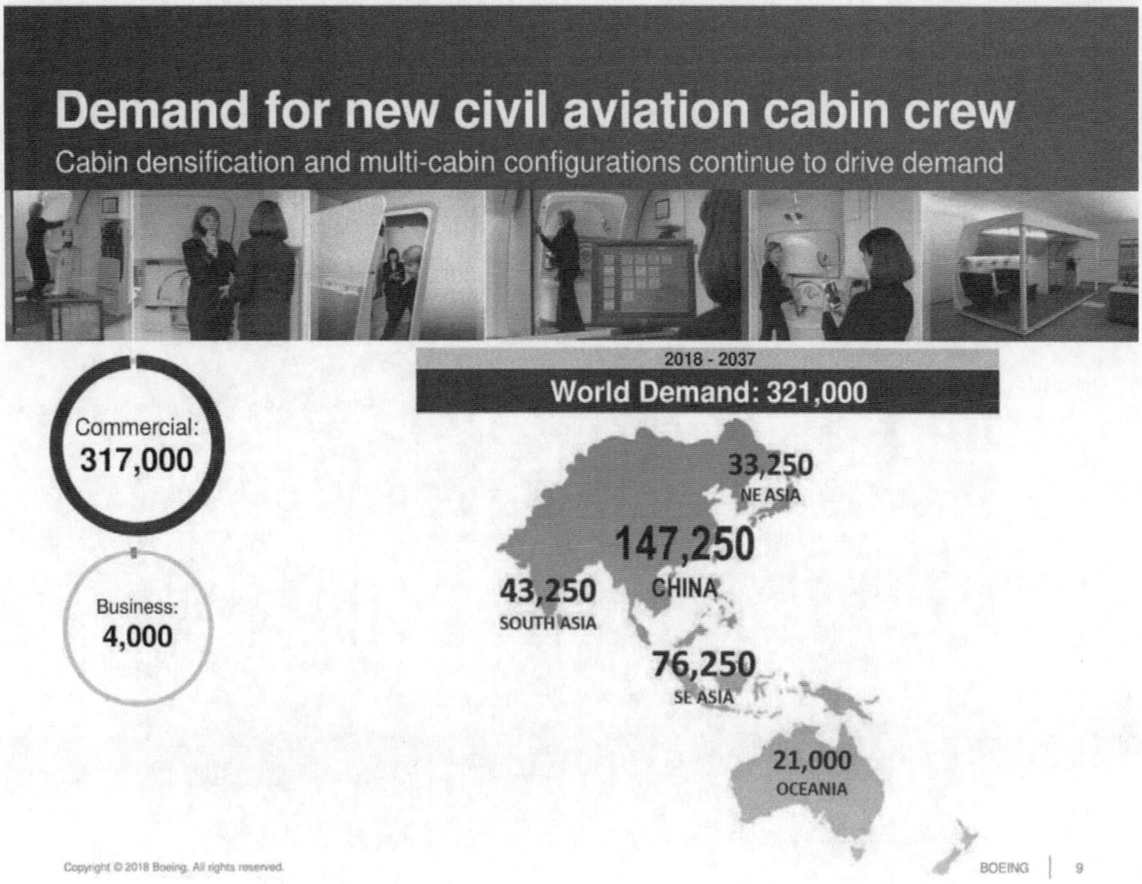
Demand for new civil aviation cabin crew
Cabin densification and multi-cabin configurations continue to drive demand
Commercial:
317,000
Business:
4,000
2018 - 2037
World Demand: 321,000
33,250
NE ASIA
147,250
CHINA
43,250
SOUTH ASIA
76,250
SE ASIA
21,000
OCEANIA
Copyright © 2018 Boeing. All rights reserved.
BOEING | 9

Boeing Global Services Training Worldwide
Canada
Seattle
Mountain Home AFB
Denver
Mesa
St. Louis
OKC
Philadelphia
Wright Patt AFB
Seymour Johnson AFB
NAS Jacksonville
Miami
Mexico City
Buenos Aires
Argentina
Finland
Moscow
RAF Lakenheath
BDUK
Gatwick
Netherlands
NATO
Belgium
Germany
Spain
Italy
Istanbul
Turkey
Greece
Israel
Morocco
Casablanca
Libya
Egypt
Iraq
Kuwait
UAE
Oman
Saudi Arabia
Afghanistan
India
Korea
Incheon/Gimpo
Seoul
Shanghai
Japan
Tokyo
Taiwan
Thailand
Singapore
Malaysia
Australia
RAAF Amberley
Brisbane
RAAF Williamtown
Melbourne
Key :
Development Offices
Commercial Training Campuses
Pro Forma Commercial - Flight Training Campuses
Pro Forma Commercial - Maintenance Training Campus
Pro Forma Commercial - Training Campus for 787 Only
USG Defense Training Sites
Defense Training Sites
Full Corporate Resourcing for Training Around the World
Copyright © 2017 Boeing. All rights reserved.
Boeing Global Services

What is Boeing Training & Professional Services doing?

Inspiring and developing tomorrow's workforce

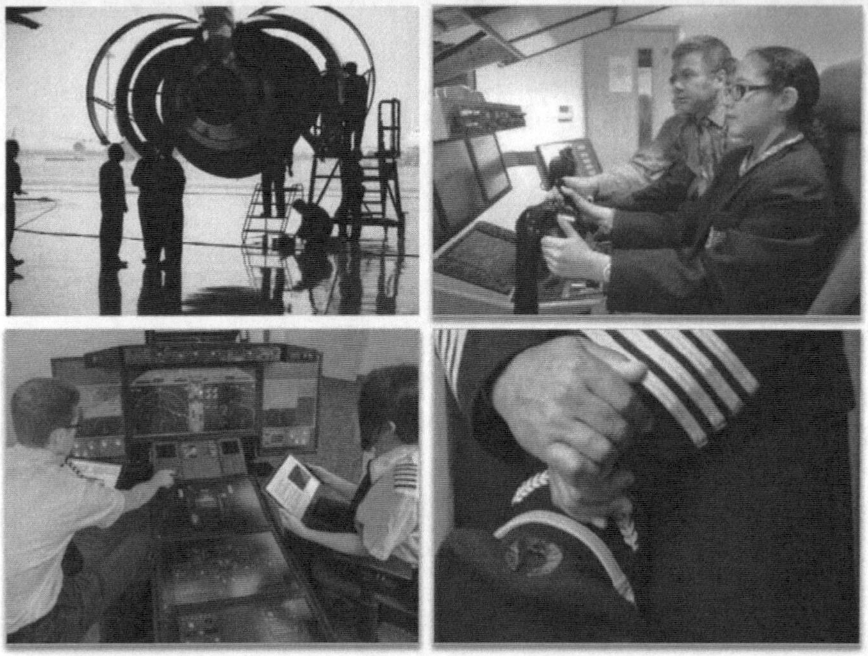

BOEING

Gaming in Cabin Crew CRM Training

Norman MacLeod
Cathay Pacific Airways

What is a game?

- An activity that captures aspects of the real world
- An environment that provides for behaviour rehearsal
- A method for reinforcing learning

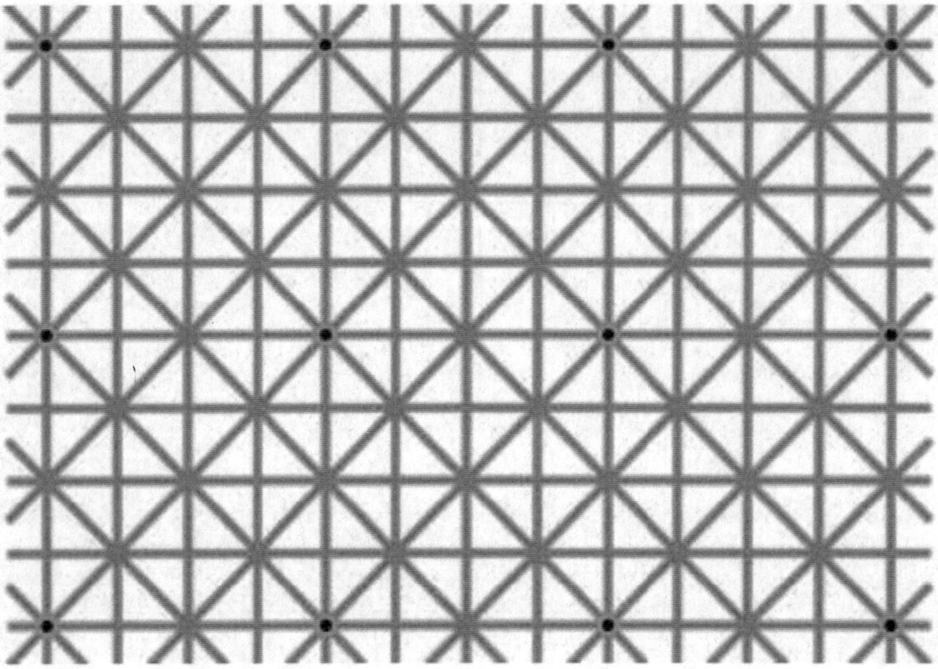

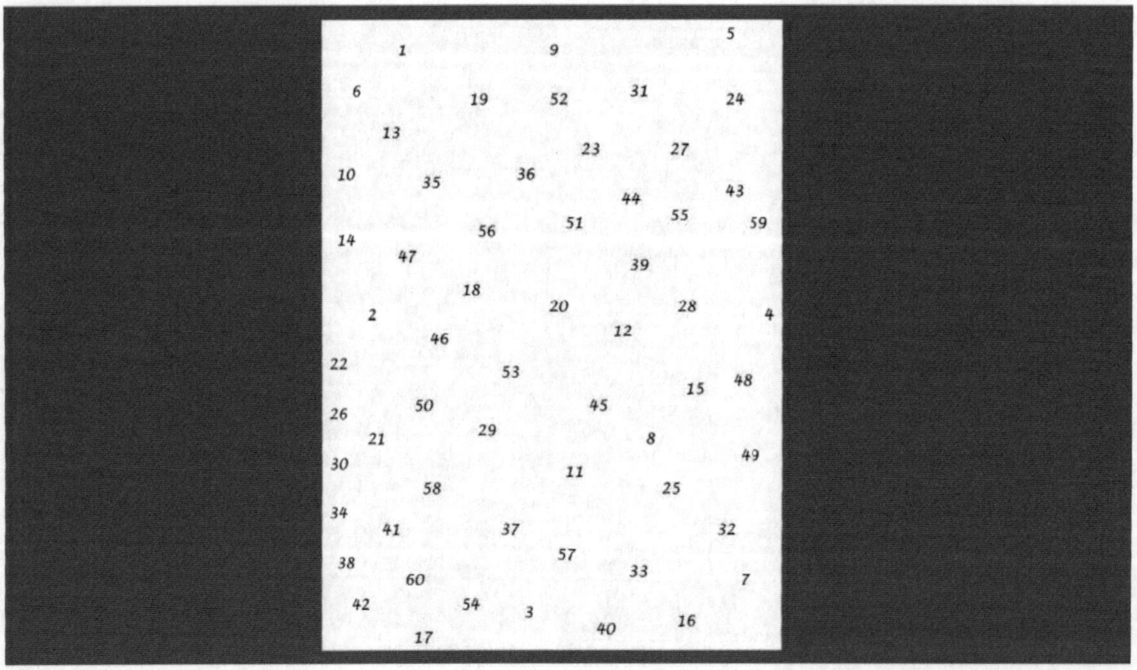
1
9
5
6
19
52
31
24
13
23
27
10
35
36
44
43
55
56
51
59
14
47
39
18
20
28
4
2
46
12
22
53
48
15
45
50
26
29
21
8
49
30
11
58
25
34
41
37
32
57
38
33
60
7
42
54
3
16
40
17

Joint Training Non-normal Situations

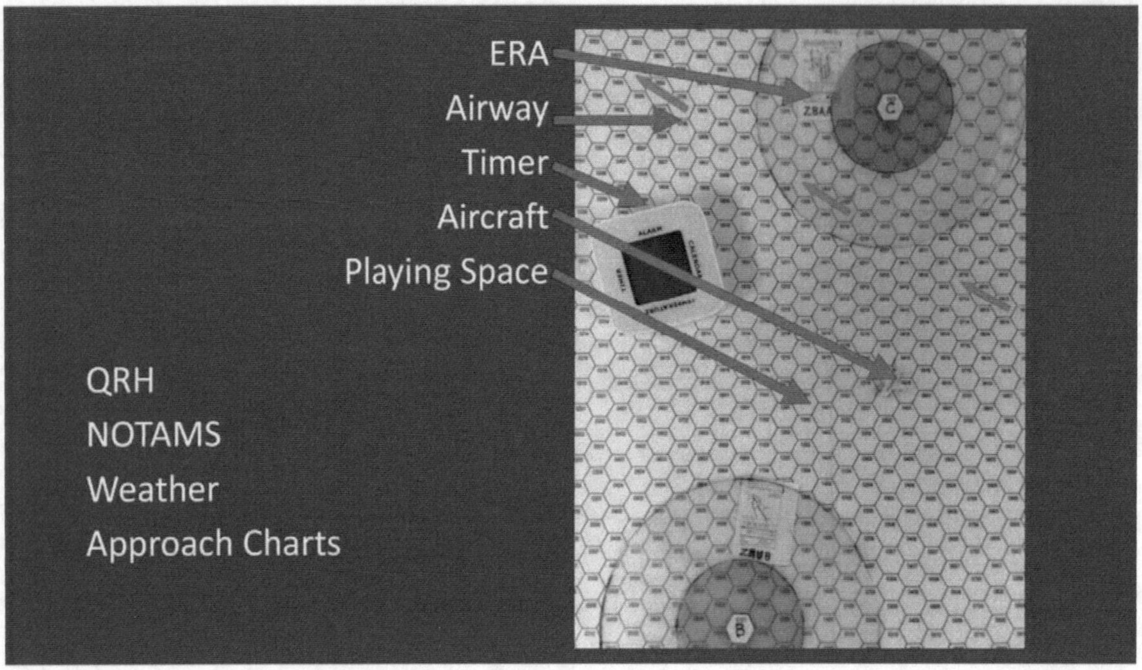
ERA
Airway
Timer
Aircraft
Playing Space
QRH
NOTAMS
Weather
Approach Charts

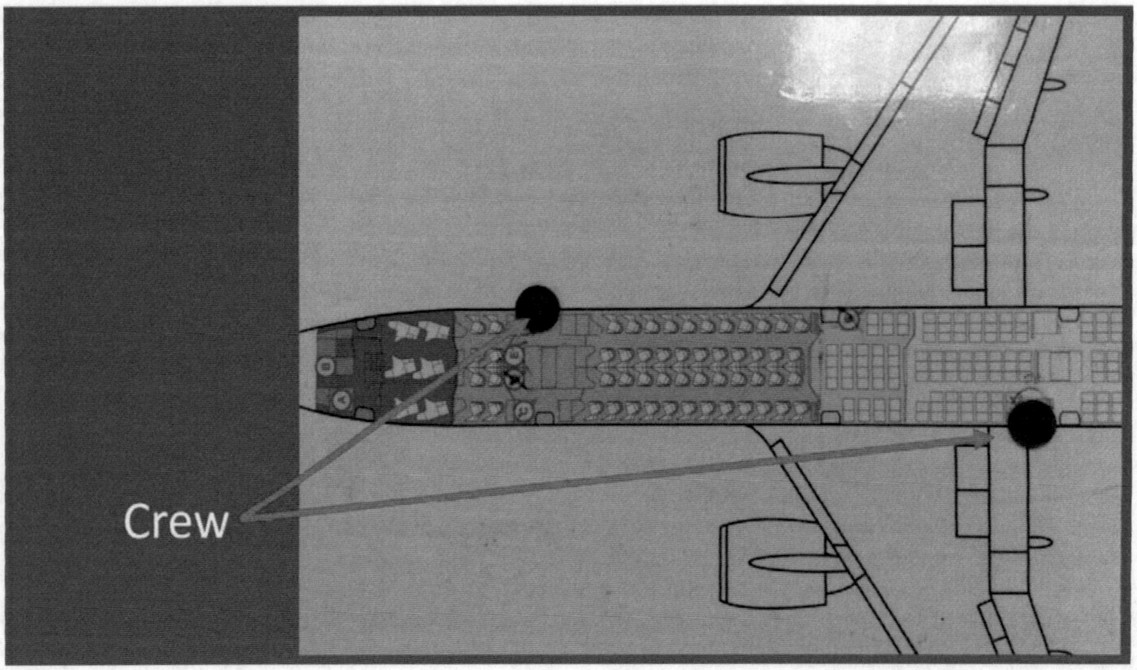
Crew

Flight Deck Door

Normal Operations

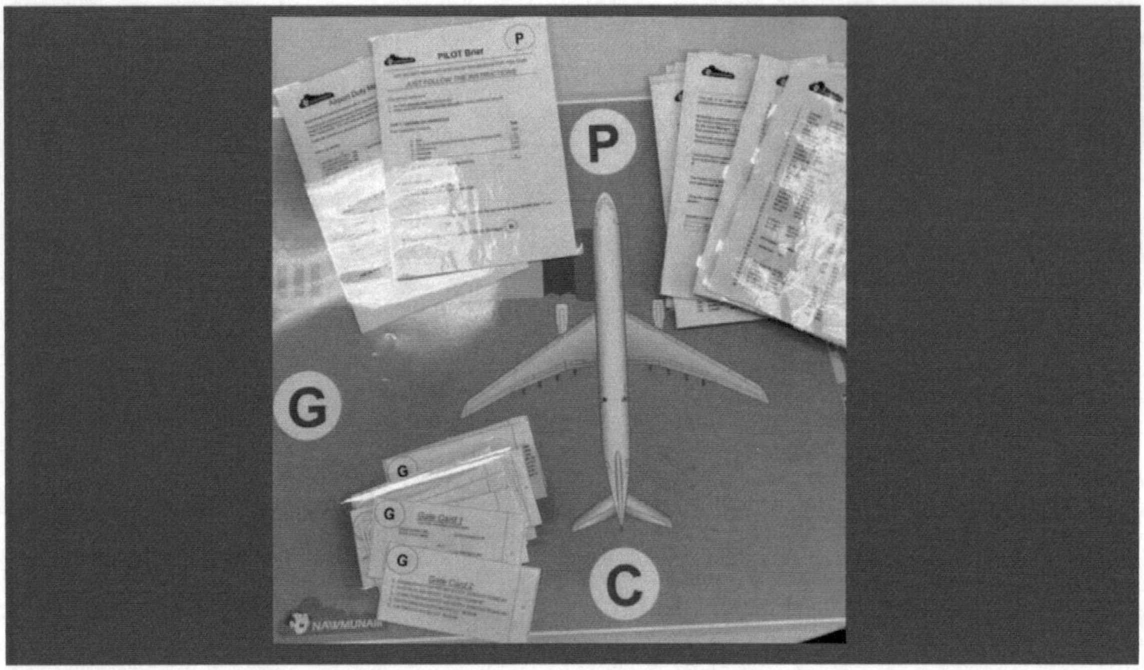
PILOT Brief
P
G
C

Gate
Pilots
Cabin
Ramp

Benefits of games?

- Engaging and motivating
- Generates emotions and 'experience'
- Allows for comparisons between performances
- Creates real linkages to real world

Thank you for your attention

Norman MacLeod
Human Factors Manager
Cathay Pacific Airways

BE WHAT YOU WANT TO BE

Next Generation of ~~CRM~~ Human Factors Training, Education & Practice!

Associate Professor Yvonne Toft
Discipline Lead: Human Factors Engineering
Head of Department: Transport, Emergency & Safety Sciences
Central Queensland University

CQUniversity AUSTRALIA

Not an aviator but rather, a pilot of complex socio-technical systems design …

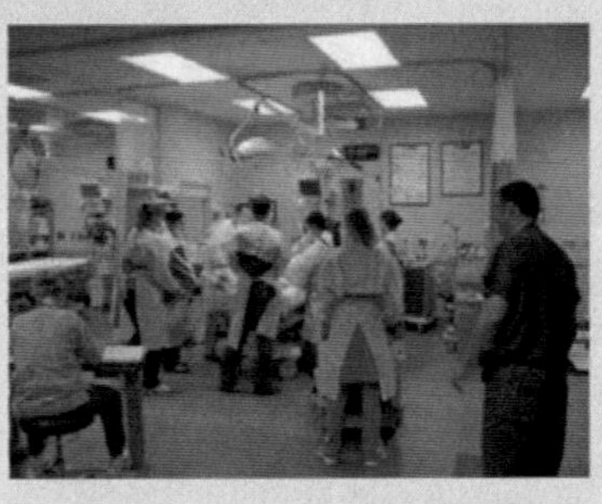

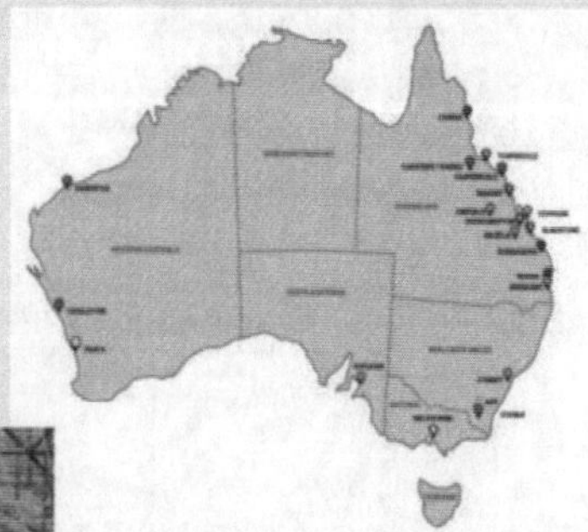

CQUniversity AUSTRALIA

Why Next Gen? – what is wrong with the way we currently think about Human Factors???

- Just telling us about errors will not actually stop us making one
- Impressive human factors in engineering input at front end of cabin design but little evidence of cabin crew input as the client and user the interacts most frequently with the system
- 'Shopping centre' design issues
- Interoperability 'rental car' as fleets are changed and managed
- Limited human factors investigation of events, data flawed by the lens of the investigator then poor analysis
- Events hiding in data disguised as other types of 'errors'
- Poor data limits the opportunity to learn from current analysis methods

CQUniversity AUSTRALIA

What is Human Factors?

*Ergonomics (or human factors) is the scientific discipline concerned with the understanding of interactions among humans and other elements of a system, and the profession that applies theory, principles, data and methods to **design** in order to optimize human well-being and overall system performance.*

Practitioners of ergonomics and ergonomists contribute to the design and evaluation of tasks, jobs, products, environments and systems in order to make them compatible with the needs, abilities and limitations of people.

http://www.iea.cc/whats/index.html

CQUniversity AUSTRALIA

It is all about the 'lens' we wear ...

CQUniversity AUSTRALIA

... and understanding

CQUniversity AUSTRALIA

... really 'getting' ...

In 1487

- an English Coroner ordered that a cliff be fenced off because it collapsed and caused a death.

In 1990

- at a Melbourne beachside suburb a cliff collapsed and killed a 3 year old child (Coroner's recommendations).

Again in 1990

- in country Victoria a cliff collapsed during a school abseiling excursion and course. Two girls were killed (Coroner's recommendations)

In 1995

- at Gracetown, WA, a cliff collapsed on a beach, killing 9 (Coroner's recommendations)

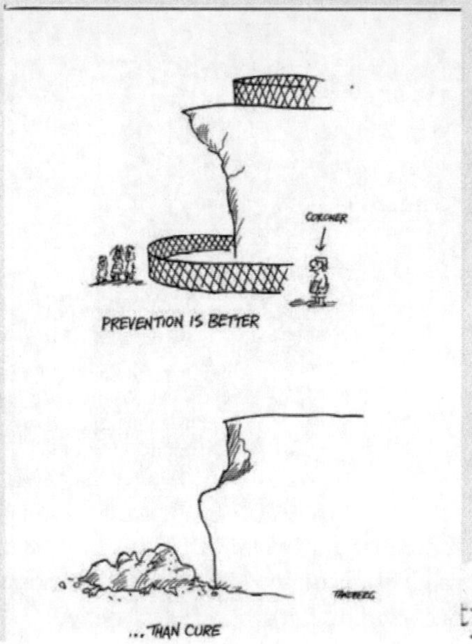

... that after 700 years of 'learning' we still make the same mistakes ...

BELLARINE Echo

A 15ha property in Hardings Rd, Wallington, sold at auction for $1.2 million on Saturday.

MONDAY: CLIFF-FACE steps at the Bird Rock end of Jan Juc will remain permanently closed following a risk assessment. The assessment was sparked by last November's collapse of an 80m section of cliff which trapped an 18-year-old youth.

advertiser.c
Editorial:
newsecho@
advertiser.c
Postal:
PO Box 113
Classified:
131 585
or **fax** 5227

We are sometimes very slow learners!

A 1950 and a 1994 photograph of a tractor rollover in country Victoria (both incidents occurred on flat ground). No rollover protection (bars) fitted.

Farming is one of the most dangerous occupations in this country, crippling and maiming thousands every year and leaving others dead. But it's only now that officialdom is waking up to the problem, reports John Stapleton

THE KILLING FIELDS

CQUniversity AUSTRALIA

But what about those pesky humans?

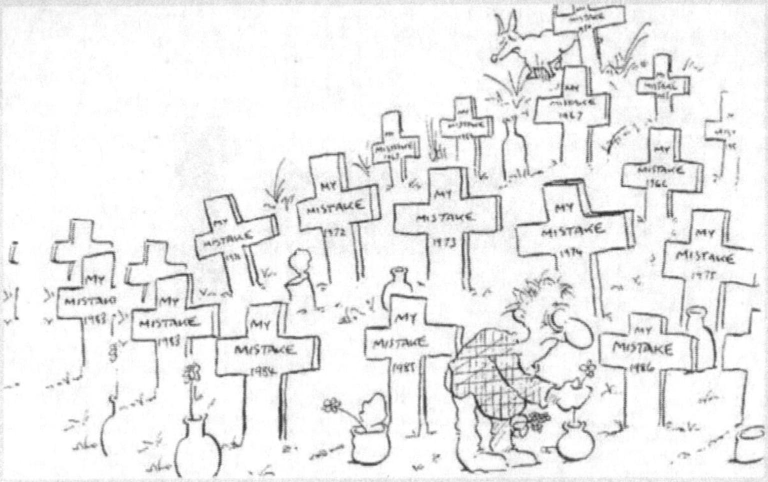

CQUniversity AUSTRALIA

We need to understand what made sense to them in real time ...

- *To understand the **relationships** between the technical and social elements of the system ... where the balance worked or did not in the quest for productivity!*
- *To understand **why**!*
- *To **learn**!!!*

CQUniversity AUSTRALIA

We need to understand human factors illusions!!!

Reality is merely an illusion, albeit a persistent one ...

CQUniversity AUSTRALIA

Illusion 1

Are we solving the
***'REAL' problem**?*

We can't prevent accidents …

... if we only look for the obvious!!!

Capricorn Coast – 0407 039 977 — We live here — 7

Simple human error was the cause of Monday's plane crash

Distracted pilot left fuel cap off

A little help from

SPECIAL REPORT: Why there's so much trouble in the air

A plane crashes every fortnight

He left the fuel cap off

...but I share the blame, says chatty passenger

It all depends on the lens we bring ...

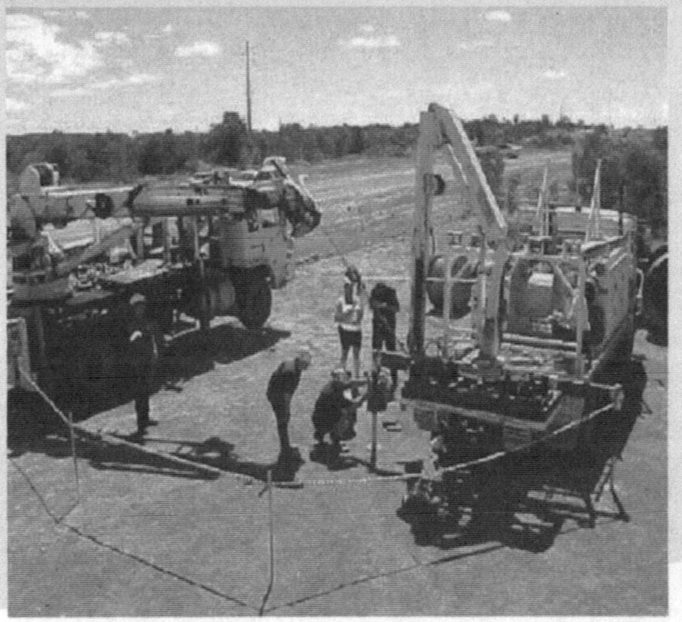

A simple case of human error …

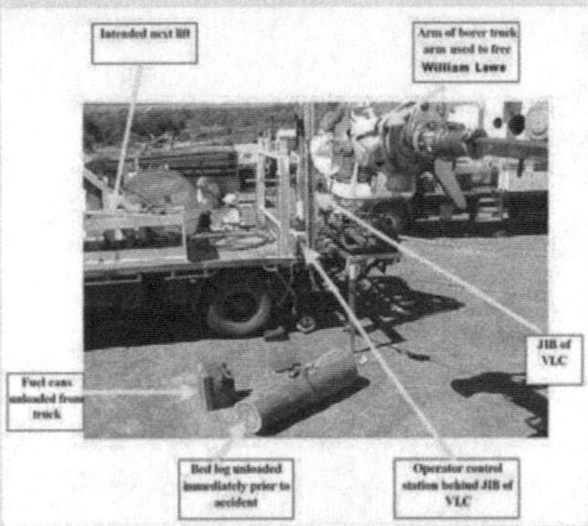

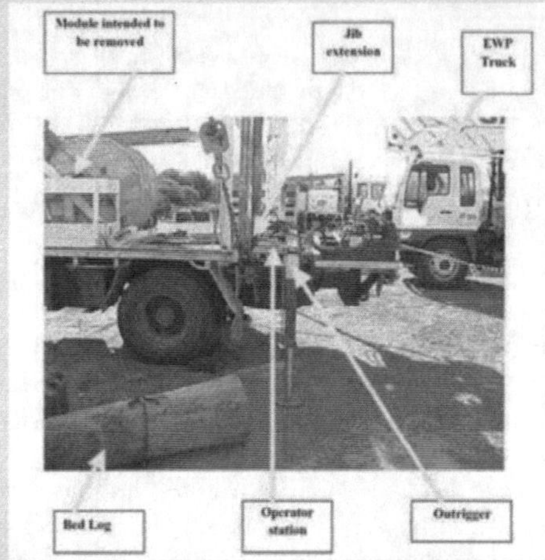

CQUniversity AUSTRALIA

… wrong procedures or wrong action??!

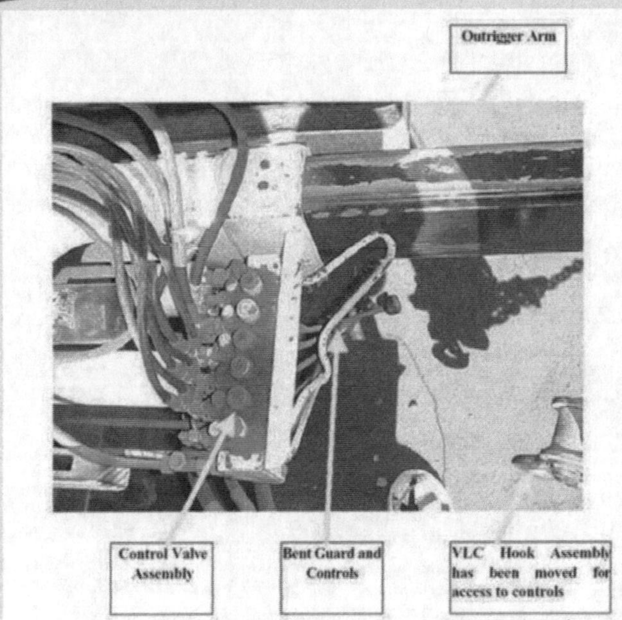

CQUniversity AUSTRALIA

Illusion 2

Are we designing for the ***'REAL' users*** *who interact with our system across the system life cycle?*

CQUniversity AUSTRALIA

Can a designer understand those who will interact with their system ...

Test of the Hamburg Chair Technical Session Tomorrow (1/2)

- Hands-on at aircraft and user-test
- Observation, Questioning and Expert Judgements
- Feedback from
  - Wheelchair user
  - Airlines
  - Experts

Observation Sheet Human Factors
On-board Wheelchair

Date: Observer:

Type of Aircraft: Type of LAV: Location:

Space in front of lavatory:

Transfer Type: PIT ☐ PAT ☐ IT ☐ DT ☐

Wheelchair Type:

Remarks Test Environment:

Summary Test Conclusion:

Part 1:

Topic	Test Criteria	Observations	Comments	Judgement
Cabin Crew to prepare Hamburg Chair for PRM transport in cabin	1) Hamburg Chair handling safe 2) Perception of Hamburg Chair			
Transfer of PRM from seat to Hamburg Chair	1) Safe transfer			

Page 2

AIRBUS

Accessible Aircraft Cabin - ECGC6 - Ref. PR1614300 - Issue 1 17. August 2016

Test of he Hamburg Chair Technical Session Tomorrow (2/2)

NR	Task description	Criteria
1	Cabin Crew to prepare Hamburg Chair for PRM transport in cabin	Hamburg Chair handling safe & perception of Hamburg Chair
2	Transfer of PRM from seat to Hamburg Chair	safe transfer
3	Cabin Crew to transport PRM to lavatory	safe transfer
4	Cabin Crew to prepare lavatory for access	sufficient space in front of lavatory & privacy
5a	Independent use of lavatory by PRM	accessibility, privacy, suitable interior
5b	Use of lavatory by PRM with assistant	accessibility, privacy, suitable interior
6	Prepare Hamburg Chair and user for transfer back to cabin seat	Hamburg Chair handling safe & perception of Hamburg Chair
7	Cabin Crew to transport PRM back to cabin seat	safe transfer

Page 10

AIRBUS

... and really wear THEIR lens?

CQUniversity AUSTRALIA

CQUniversity
AUSTRALIA

Our ability to create real problems for those who interact with our systems astounds ...

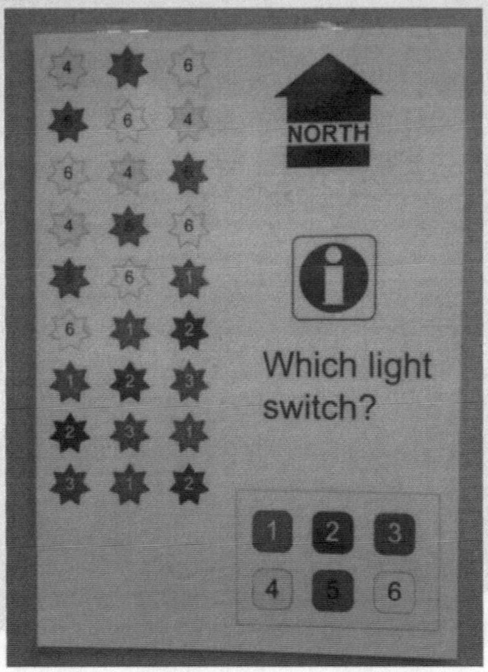

CQUniversity
AUSTRALIA

Illusion 3

Have we found a
***'REAL' solution**?*

Socio-technical systems are deceivingly complex ...

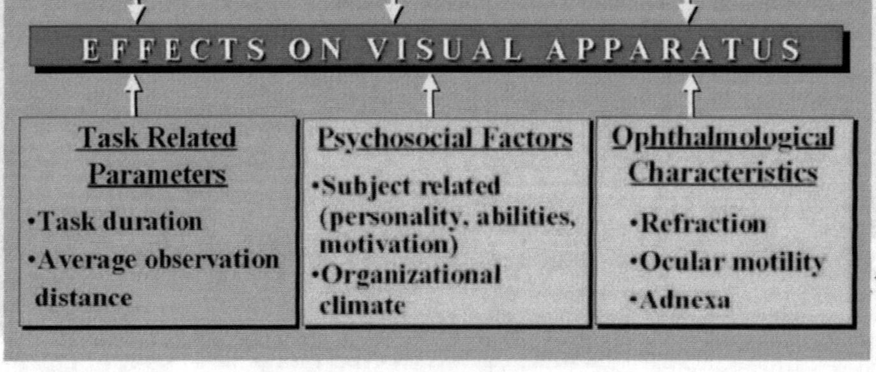

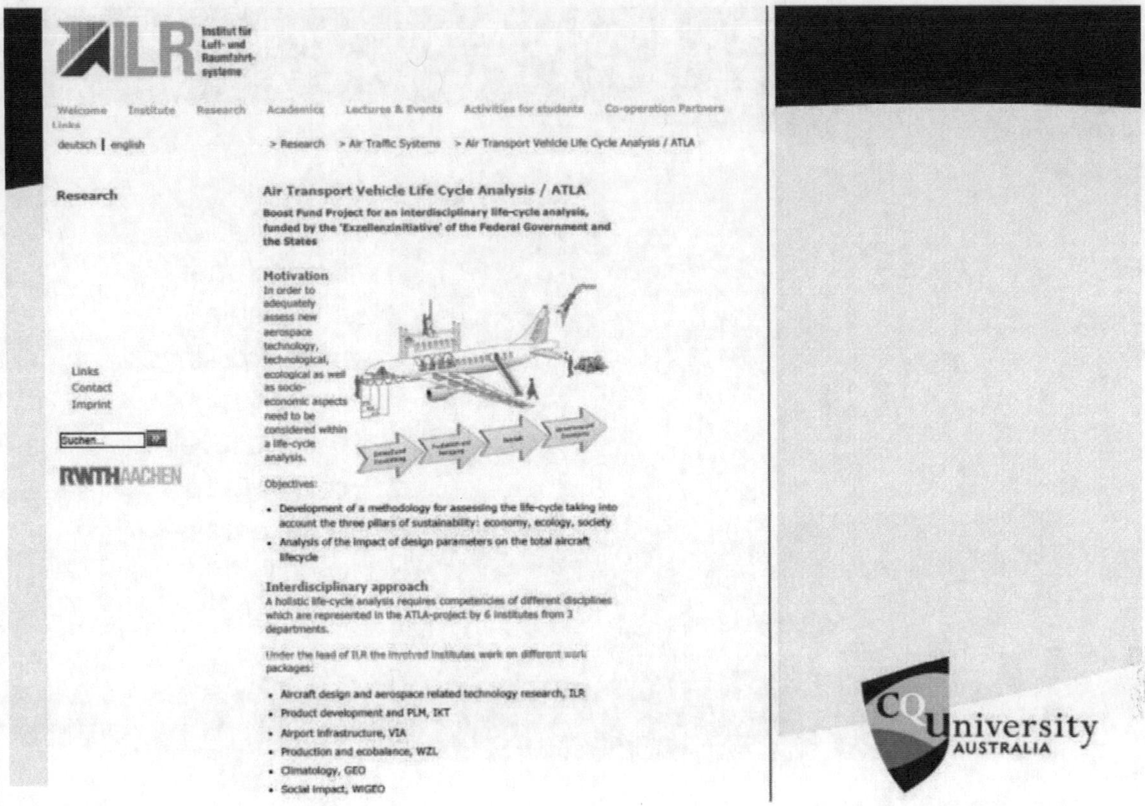
ILR
Institut für
Luft- und
Raumfahrt-
systeme
Welcome Institute Research Academics Lectures & Events Activities for students Co-operation Partners
Links
deutsch | english
> Research > Air Traffic Systems > Air Transport Vehicle Life Cycle Analysis / ATLA
Research
Air Transport Vehicle Life Cycle Analysis / ATLA
Boost Fund Project for an interdisciplinary life-cycle analysis, funded by the 'Exzellenzinitiative' of the Federal Government and the States
Motivation
In order to adequately assess new aerospace technology, technological, ecological as well as socio-economic aspects need to be considered within a life-cycle analysis.
Links
Contact
Imprint
Suchen...
RWTH AACHEN
Objectives:
Development of a methodology for assessing the life-cycle taking into account the three pillars of sustainability: economy, ecology, society
Analysis of the impact of design parameters on the total aircraft lifecycle
Interdisciplinary approach
A holistic life-cycle analysis requires competencies of different disciplines which are represented in the ATLA-project by 6 institutes from 3 departments.
Under the lead of ILR the involved institutes work on different work packages:
Aircraft design and aerospace related technology research, ILR
Product development and PLM, IKT
Airport infrastructure, VIA
Production and ecobalance, WZL
Climatology, GEO
Social impact, WIGEO
CQUniversity
AUSTRALIA

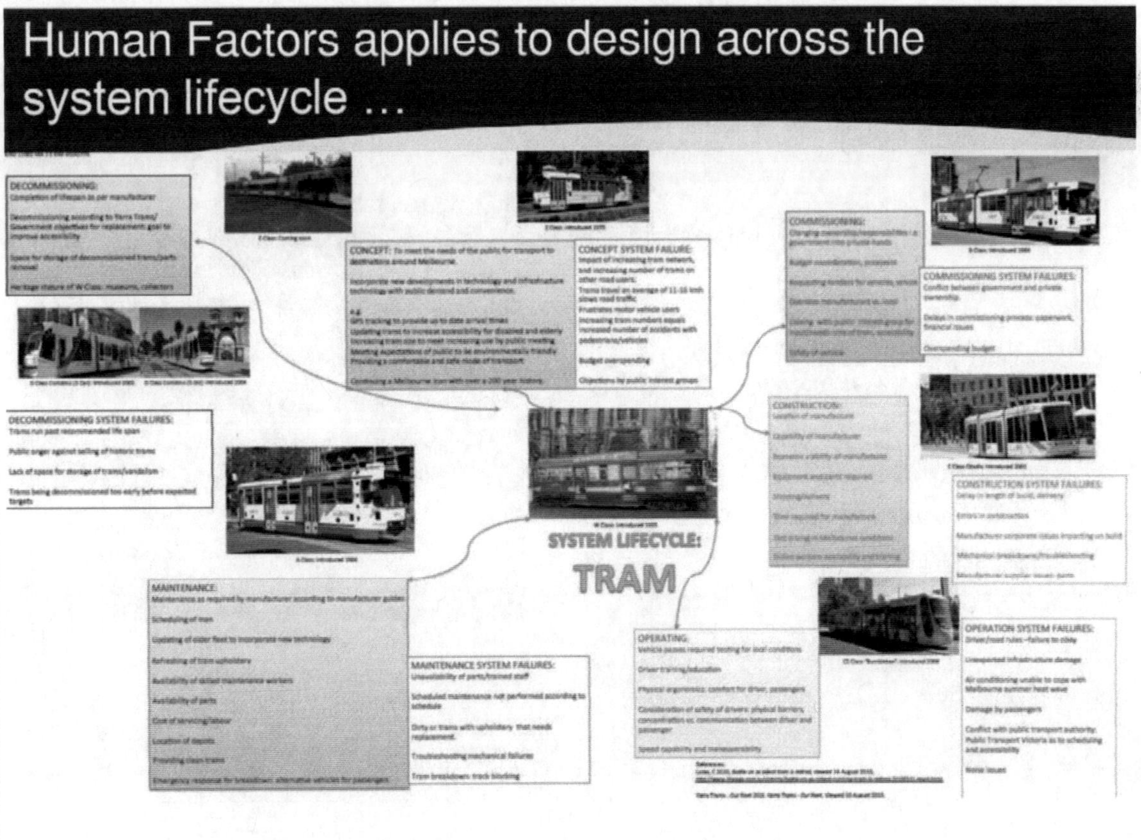
Human Factors applies to design across the system lifecycle …
DECOMMISSIONING:
DECOMMISSIONING SYSTEM FAILURES:
Trams run past recommended life span
Public anger against selling of historic trams
Lack of space for storage of trams/vandalism
Trams being decommissioned too early before expected targets
CONCEPT: To meet the needs of the public for transport to destinations around Melbourne.
CONCEPT SYSTEM FAILURE:
Budget overspending
Objections by public interest groups
COMMISSIONING:
COMMISSIONING SYSTEM FAILURES:
Conflict between government and private ownership
Delays in commissioning process: paperwork, financial issues
Overspending budget
CONSTRUCTION:
CONSTRUCTION SYSTEM FAILURES:
SYSTEM LIFECYCLE:
TRAM
MAINTENANCE:
Maintenance as required by manufacturer according to manufacturer guides
MAINTENANCE SYSTEM FAILURES:
Troubleshooting mechanical failures
Tram breakdown: track blocking
OPERATING:
Driver training/education
Speed capability and manoeuvrability
OPERATION SYSTEM FAILURES:
Driver/road rules - failure to obey
Unexpected infrastructure damage
Air conditioning unable to cope with Melbourne summer heat wave
Damage by passengers
Noise issues

... and can't achieve this when we don't question elusive assumptions!

Design brief - affordable level crossings i.e. redesign passive level crossings to reduce train-vehicle collisions ...

CQUniversity AUSTRALIA

Wullems (circa 2012)

When questioned about his design assumptions ...

"my first assumption had to be that people would use the crossing properly otherwise how would I be able to assign a risk score ... "

CQUniversity AUSTRALIA

Illusion 4

*Are our decisions based on **'REAL' evidence**?*

CQUniversity AUSTRALIA

Investigation of human error is only as useful as the lens we wear while investigating ...

Even unsophisticated HOLISTIC models can begin to illustrate the complexity that underpins human error across the system lifecycle and helps us identify the genesis of the errors in our systems

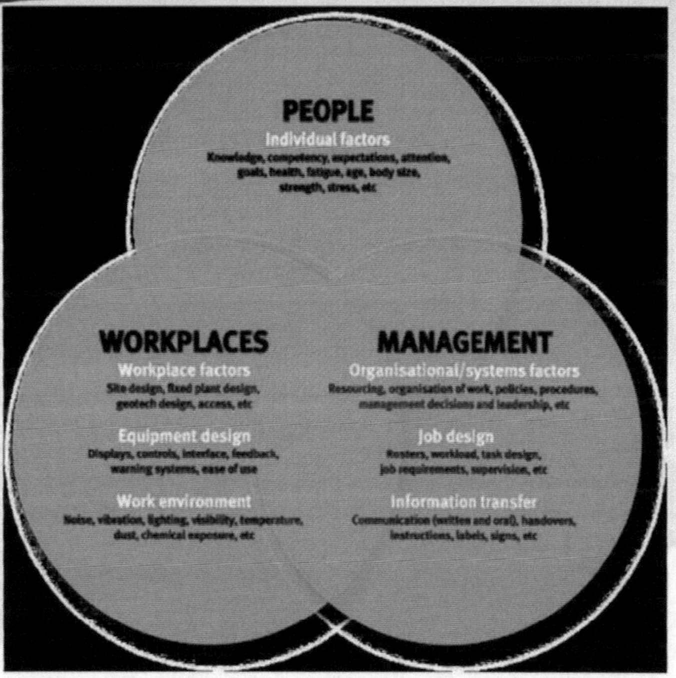

However, if we complete air safety investigations with a linear lens …

1950 – Bird & Germain – Loss Causation Model

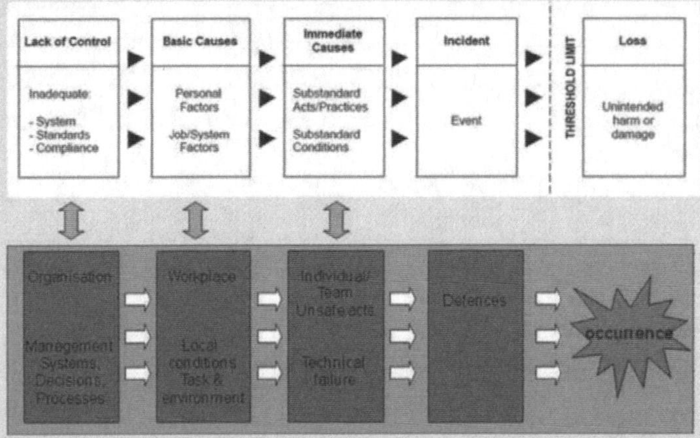

1990 – Reason – Model of Organisational Accidents

… and classifications, for example, HFACS …

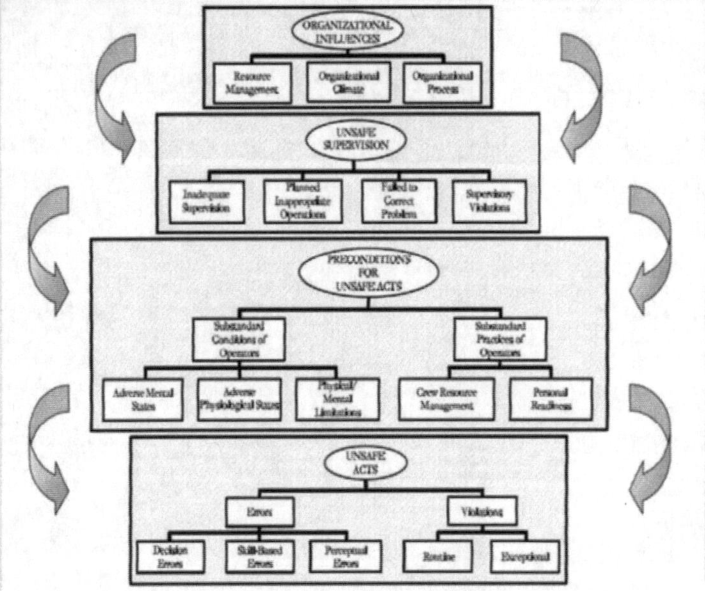

… we can fail to see the obvious and the hidden no matter the flavour of the month ….

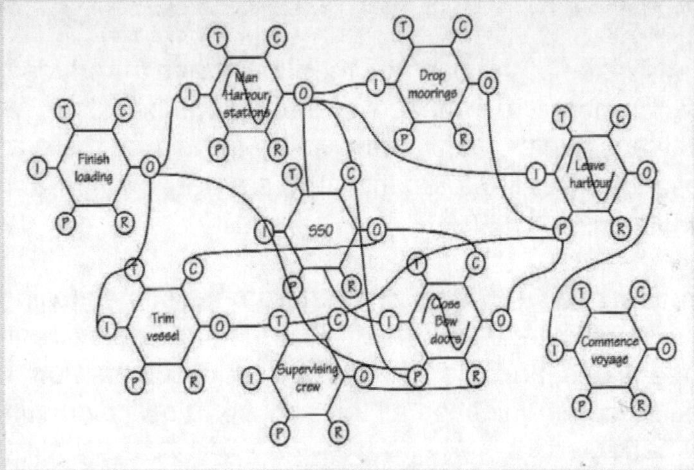

Hollnagel: Functional Resonance Accident Method

CQUniversity AUSTRALIA

The obvious …

CQUniversity AUSTRALIA

The hidden ...

Fatigue is a serious issue at the best of times, but when introduced to safety-critical environments, the impacts of fatigue can be fatal. While there have not been any accidents directly attributable to fatigue, there are many instances where fatigue has led to an incident (a safety event that has not led to loss of life or personal injury).

Cabin crew have described incidents ranging from as little as falling asleep in jump seats to major incidents involving unintentional deployment of emergency escape slides, and pilots have reported incidents from missed radio calls and failed communication all the way up to miscalculating fuel figures and flight paths.

Over 500 incidents in 2018!

CQUniversity AUSTRALIA

Classifying an error out of context will do NOTHING to help us learn!

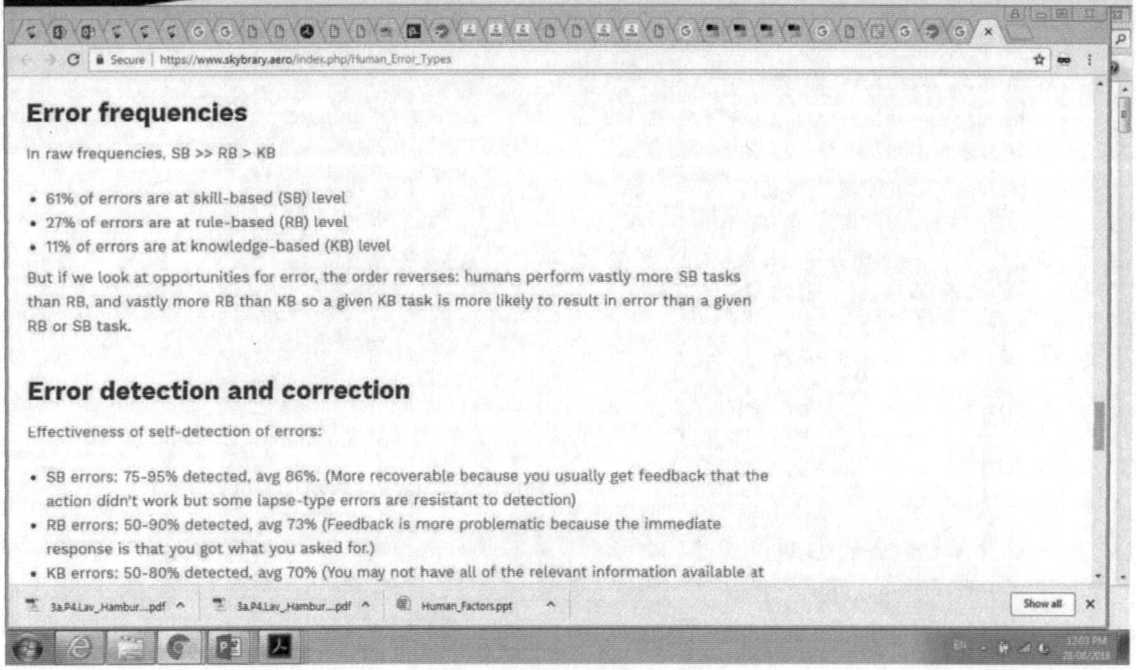

Error frequencies

In raw frequencies, SB >> RB > KB

- 61% of errors are at skill-based (SB) level
- 27% of errors are at rule-based (RB) level
- 11% of errors are at knowledge-based (KB) level

But if we look at opportunities for error, the order reverses: humans perform vastly more SB tasks than RB, and vastly more RB than KB so a given KB task is more likely to result in error than a given RB or SB task.

Error detection and correction

Effectiveness of self-detection of errors:

- SB errors: 75-95% detected, avg 86%. (More recoverable because you usually get feedback that the action didn't work but some lapse-type errors are resistant to detection)
- RB errors: 50-90% detected, avg 73% (Feedback is more problematic because the immediate response is that you got what you asked for.)
- KB errors: 50-80% detected, avg 70% (You may not have all of the relevant information available at

We need good data AND good analysis!

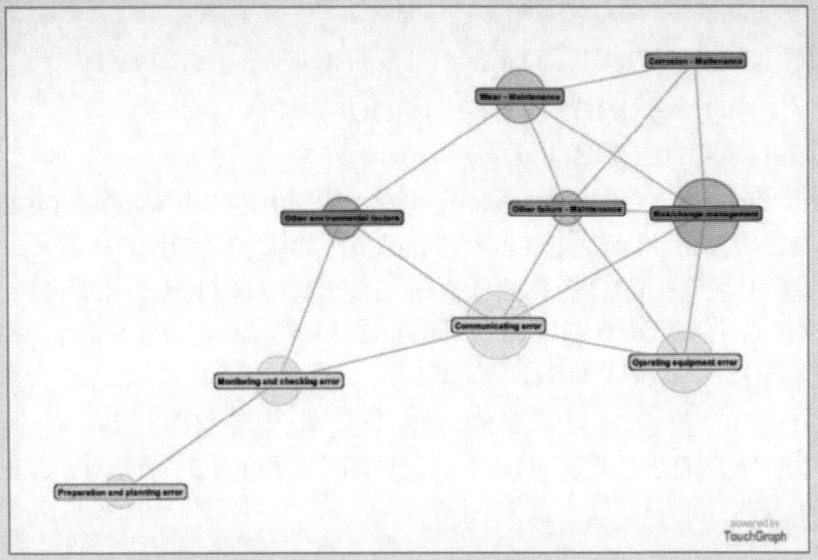

CQUniversity AUSTRALIA

We need to challenge the way we do stuff to make real change!!!

Challenge if we are

- solving the 'real' problem;
- designing for the 'real' users;
- finding the 'real' solutions; and
- basing our system designs on 'real evidence' gathered from robust investigations!

The major cause every event, we are told - Human error!

We need to use human factors more effectively …

- AS a design science that requires a transdisciplinary approach
- we will have many more opportunities to address human error across our system by tweaking the design of equipment and systems rather than by tweaking the design of people (which can't be done) and their behaviour!
- human factors in aviation is REALLY about optimising the design of complex socio-technical systems of which humans are an integral part!

CQUniversity AUSTRALIA

Complex socio-technical systems investigation and problem solving requires a transdisciplinary approach …

CQUniversity AUSTRALIA

Transport & Safety Sciences: CertIV to PhD

ACCIDENT FORENSICS & INVESTIGATION

- CertIV Rail Safety Investigation (co-badged with RISSB)
- Bachelor of Accident Forensics
- Graduate Certificate in Fatigue Risk Management
- Grad. Cert. in Advanced Accident Investigation Practice
- Graduate Diploma in Accident Investigation
- Master of Safety Science - Air Safety Investigation
- Master of Safety Science - Rail Safety Investigation
- Master of Safety Science - Road Safety Investigation
- Master of Safety Science - Industrial Safety Investigation
- Master of Advanced Safety Science Practice
- Masters by Research
- PhD

SAFETY & PUBLIC HEALTH

- CertIV in WHS
- Diploma in WHS
- Assoc. Degree in OHS
- Bachelor OHS
- Grad Cert Advanced Safety Science Practice
- Graduate Diploma in OHS
- Master of Safety Science (Major)
- Master of Advanced Safety Science Practice
- Masters by Research
- PhD

AVIATION

- Diploma in Aviation Theory
- Associate Degree in Aviation
- Bachelor of Aviation (Flight Operations)

In 2018 with:

- 200 Next Generation Accident Investigation students
- 250 Safety & Health students
- Over 2000 course (Unit) enrolments, including elective students from other programs

CQUniversity AUSTRALIA

CQU FORENSIC INVESTIGATION TEACHING LABORATORY, BUNDABERG, 2016

Dealing with cabin fires

by Robin Pijnaker

CEO - Flame Aviation

28th of August 2018

Why have cabin fires become a 'fact of life'?

- We carry more and more PED's (Personal Electronic Devices) into the cabin
- That causes a serious increase of cabin fire incidents.
- It is not a question *if* your airline will have a cabin fire incident,
  but *when* and *how often*.

The Top-3 cabin fires are:

1. Smartphone or tablet cracked in between the seats
2. Laptop overheats and ignites
3. (Cheap) power banks or chargers overheat

What are Personal Electronic Devices (PED)?

- Smartphones
- Tablets
- Chargers/power banks
- Laptops
- Headphones
- Drones
- Cameras
- E-readers

etc. etc.

If an average passengers carries two or three of these items, there are easily 500 PED's on board!

FLAME AVIATION

Why do PED's cause cabin fires?

They include lithium-ion batteries.
These are large concentrations of energy.
They can ignite if they get:

- Short-circuited
- Overheated
- Cracked (causes short-circuit)

FLAME AVIATION

Examples of cabin fires:

Indigo, Galaxy Note 2, September 2016

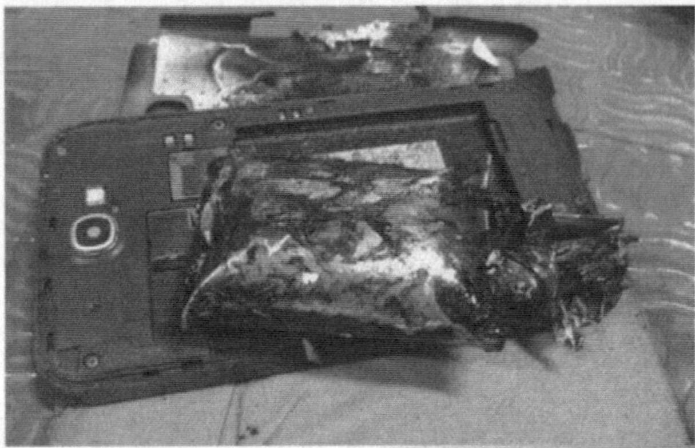

FLAME AVIATION

Examples of cabin fires:

China Airlines, Galaxy S6, December 2016

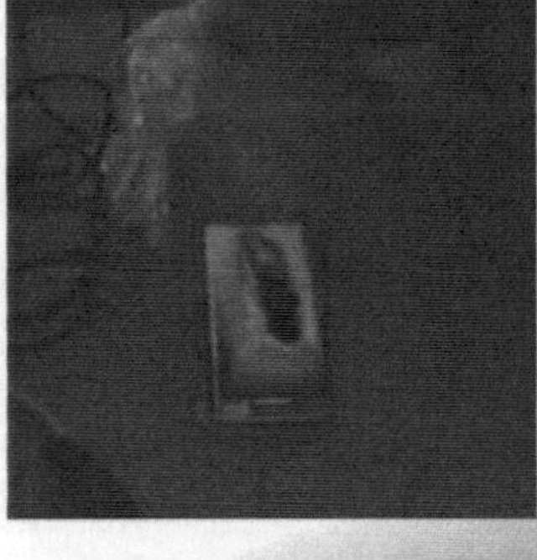

FLAME AVIATION

Examples of cabin fires:

Alaska Airlines, iPhone-6, March 2016, over the pacific ocean

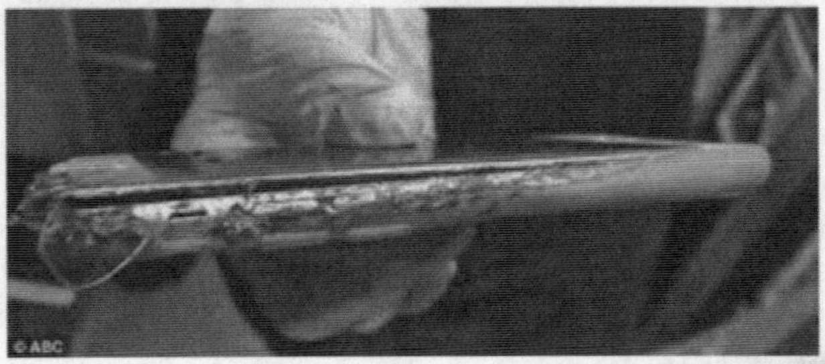

FLAME AVIATION

Examples of cabin fires:

KLM, Drone in luggage compartment, March 2015

FLAME AVIATION

What are the consequences?

Primary consequences:

- Fire
- Toxic smoke and fumes (oxygen masks cannot be sel

Secondary consequences:

- Panic
- Injury

Exploded head phone, Australia, March 2017

Does this look familiar?

..or this?

However, in an aircraft cabin:
..there is only limited space
..you might be closer to the fire
..and really feel the heat
..there might be obstructions
..smoke may reduce your visibility
..you cannot escape
..there is a buddy to help you
..the fire might be hidden or concealed
..there might be relocations or even panic
..there might be injured passengers

Question: is this training sufficient?

FLAME AVIATION

So, it's time for professional fire training...

Objective: make sure cabin crew does have the *knowledge*, *skills* and *confidence* to deal adequately with cabin fires.

1. Training program: theoretical knowledge + practical training
2. Training equipment: realistic cabin environment, real fires, real smoke, etc.
3. Procedures: train all the particular procedures for all types of cabin fires: open fires, concealed fires, lithium-ion battery fires, etc.
4. Team approach: Train as a team with an attacker, buddy and communicator
5. Scenarios: train all possible scenarios, including:
   - Use all equipment available
   - Unexpected scenarios: e.g. fire *plus* a person-on-fire
   - Make it realistic!

FLAME AVIATION

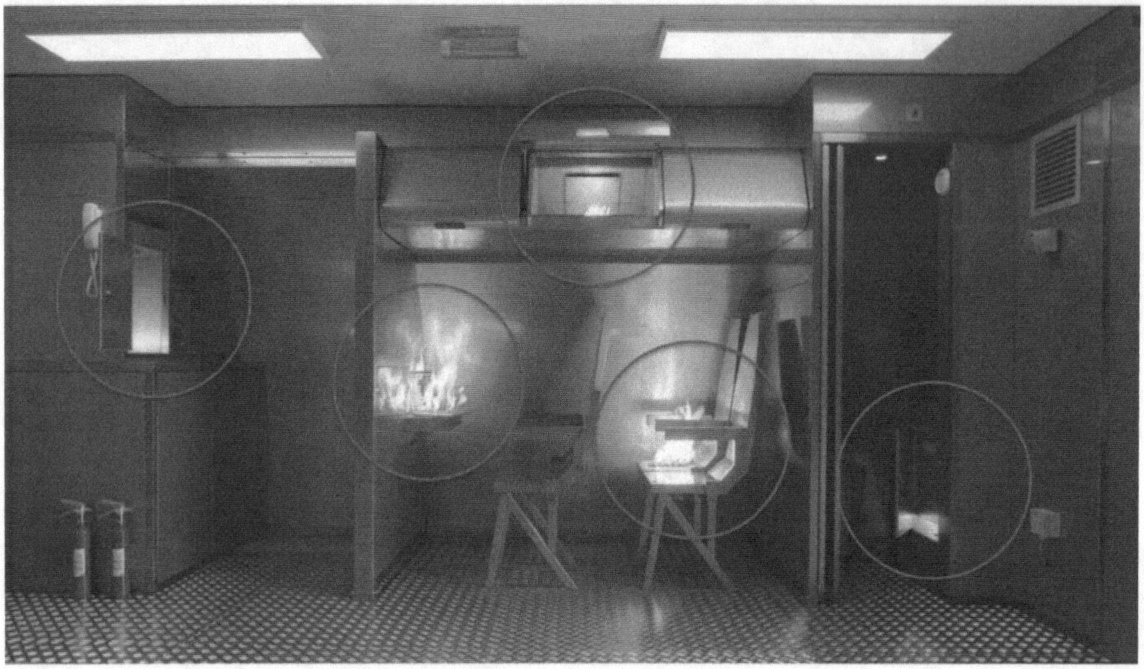

Discover smoke..

Immediate and aggressive action..

FLAME AVIATION

Get equipment immediately..

FLAME AVIATION

Personal safety first

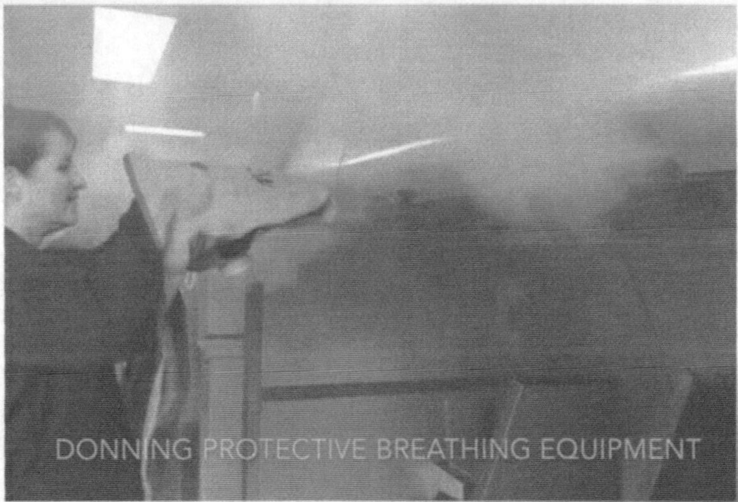

FLAME AVIATION

Buddy is standby..

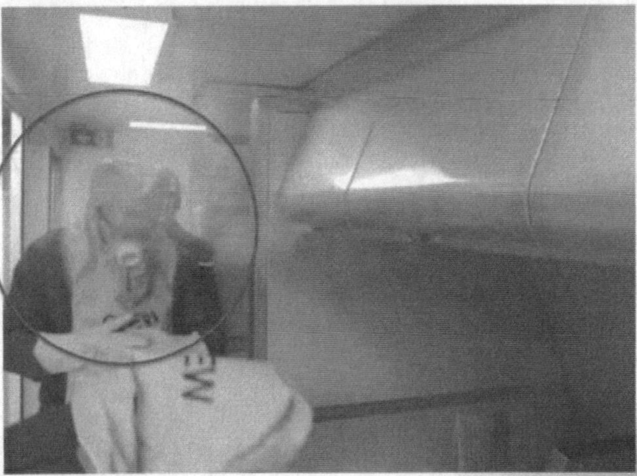

Identifying the source of the fire

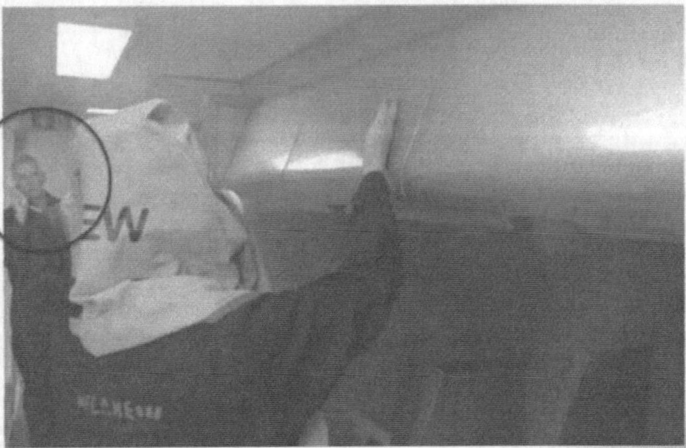

Communicator informs flight deck

Buddy opens bin just slightly

FLAME AVIATION

Close immediately after

FLAME AVIATION

Communicating with the flight deck..

- Location of the fire
- Source of the fire
- Severity of the fire, like
  - Spreading
  - Contained
  - Under control
  - Extinguished
- Number of fire extinguishers used
- Smoke conditions
- Actions taken
- Any obstructions
- Any associated hazards (e.g. toxic fumes?)
- Changes in the seating arrangement

FLAME AVIATION

Conclusions:

1. All airlines will be exposed to cabin fires, sooner or later.
2. They can be intense and have the potential to be catastrophic
3. Training programs are not always up to date
4. Realistic training equipment is not always available
5. Well-trained cabin crew is able to:
   - to neutralize the fire threat by immediate and resolute action
   - give the flight crew a clear picture of the situation
   - prevent escalation, injury, panic, etc. but maybe also aircraft diversion

All good reasons to have a closer look at your current fire training!

FLAME AVIATION

Thank you for your attention,

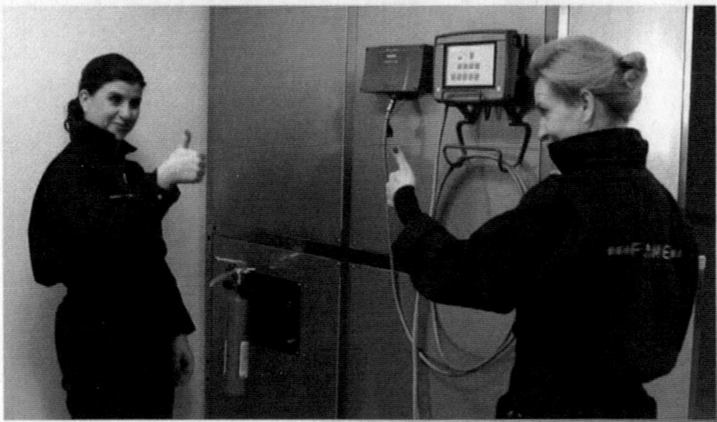

and many safe flights!

Upside and Downside of VR in Cabin Crew Training

apats2018
SINGAPORE

world of simulation

Flight attendants will increase

Passengers double from 2017 to 2036

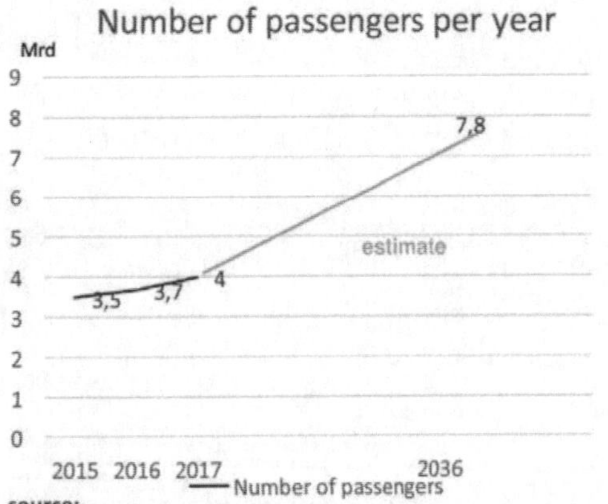

source:
https://de.statista.com/statistik/daten/studie/374860/umfrage/flugverkehr-entwicklung-passagiere-weltweit/
https://www.dfs.de/dfs_homepage/de/Presse/Publikationen/Mobilitaetsbericht2016_web.pdf

Ca. 811.000 new flight attendants to be trained

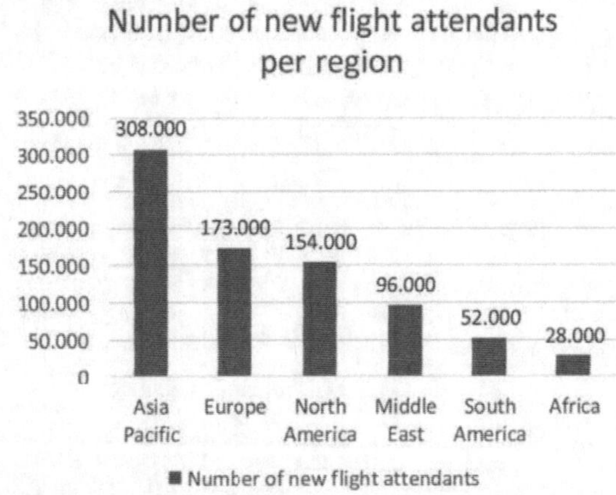

source:
https://de.statista.com/statistik/daten/studie/613308/umfrage/bedarf-an-flugbegleitern-weltweit/

world of simulation

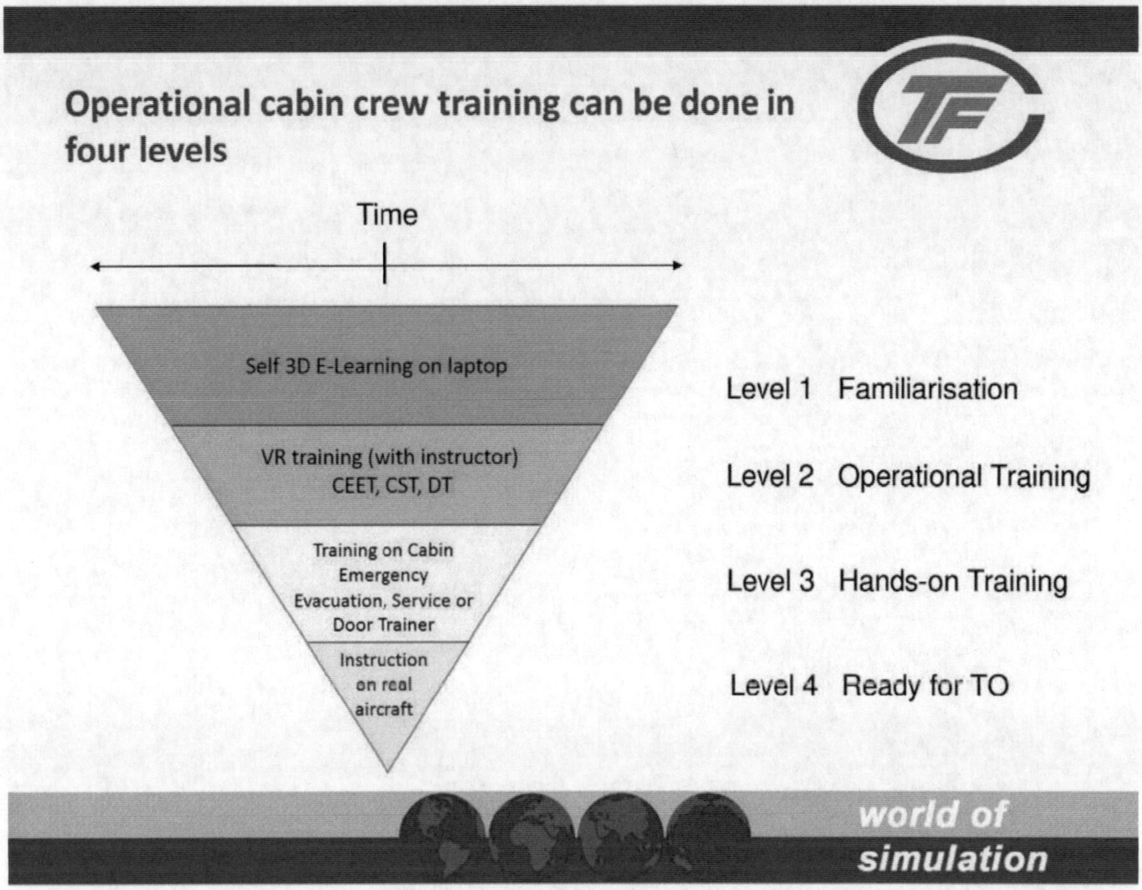
Operational cabin crew training can be done in four levels
Time
Self 3D E-Learning on laptop
VR training (with instructor)
CEET, CST, DT
Training on Cabin Emergency Evacuation, Service or Door Trainer
Instruction on real aircraft
Level 1 Familiarisation
Level 2 Operational Training
Level 3 Hands-on Training
Level 4 Ready for TO
world of simulation

ENG 1
APU
ENG 2

Effectivity of VR for cabin crew training

Study of University of Udine

48 cabin crew

24 users received instructions by traditional training

24 users received instructions by virtual reality training

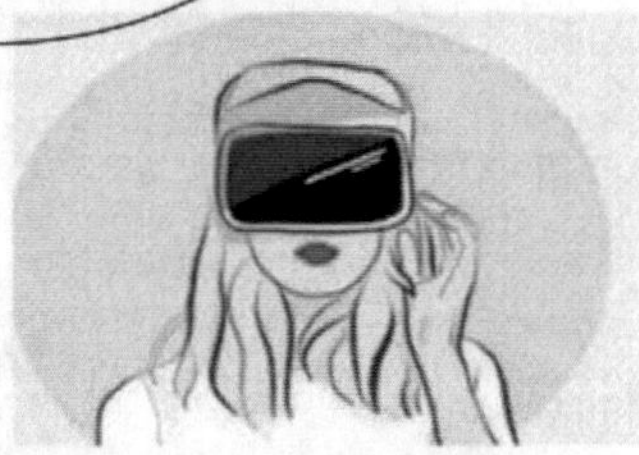

"**only 8 out of 24** members **obtained maximum score**, and scores for the remaining 16 members included also values as low as 1 and 2."

"**22 out of 24** group members **obtained maximum knowledge score (4)**, while the remaining two members obtained 3."

world of simulation

What are the advantages?

- ✓ Fast learning
- ✓ new simulation scenarios i.e. slide/raft training, bomb alert or terror attack
- ✓ Very realistic environment and sounds
- ✓ it reduces the number of mistakes by more than 50%
- ✓ Increase of confidence levels of Cabin Crew
- ✓ Replaces recurrent training (only once every 3 years recurrent training is required on real simulators)
- ✓ Saves time and resources with <u>savings of 1k€ per Cabin Crew per annum</u>

What are the disadvantages?

- ✓ No real forces
- ✓ Very forgiving
- ✓ Insuficcient for crash landings

world of simulation

BE WHAT YOU WANT TO BE

Cabin Safety Investigation:

The past & present must inform future needs

Assoc. Professor Geoff Dell, PhD

CQUniversity AUSTRALIA

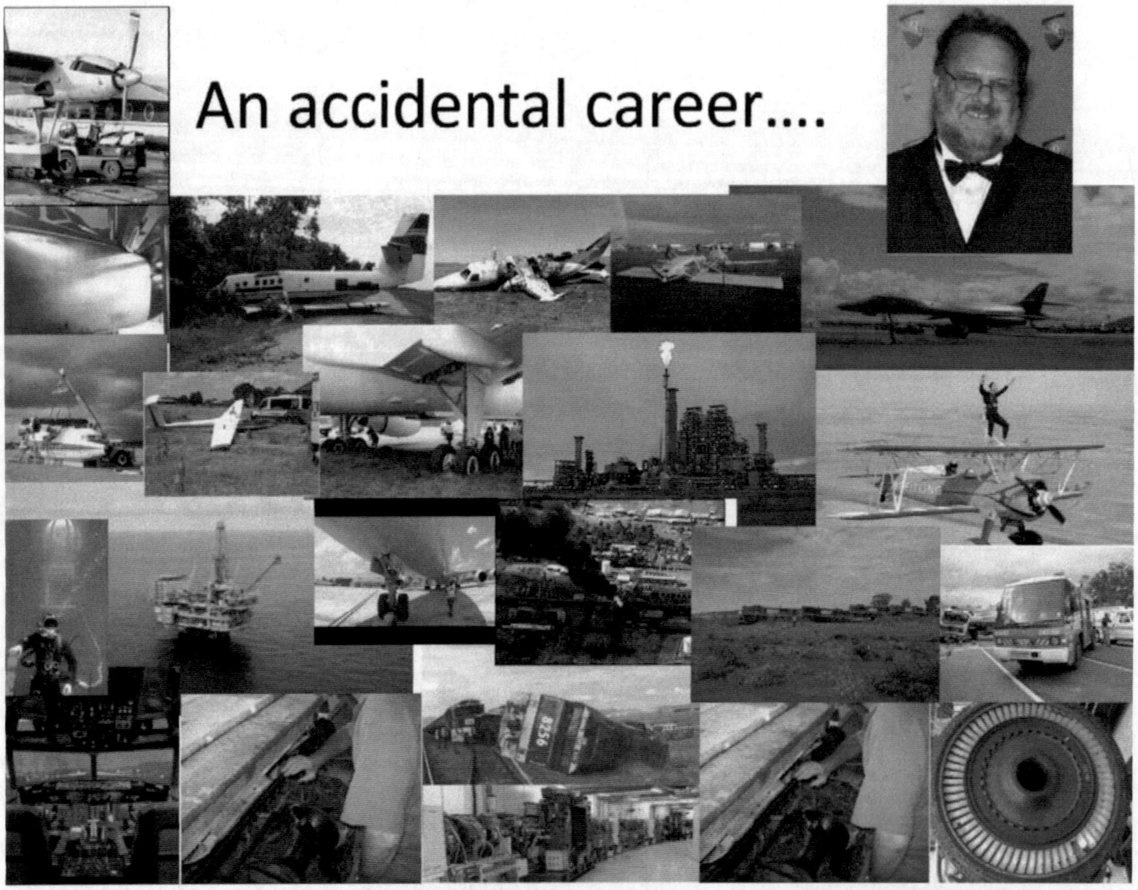

SIXTY YEARS OF JET OPERATIONS BY AUSTRALIA'S AIRLINES
NO FATAL ACCIDENTS | CIRCA 1.5 BILLION PASSENGERS CARRIED | CIRCA 15 MILLION FLIGHTS

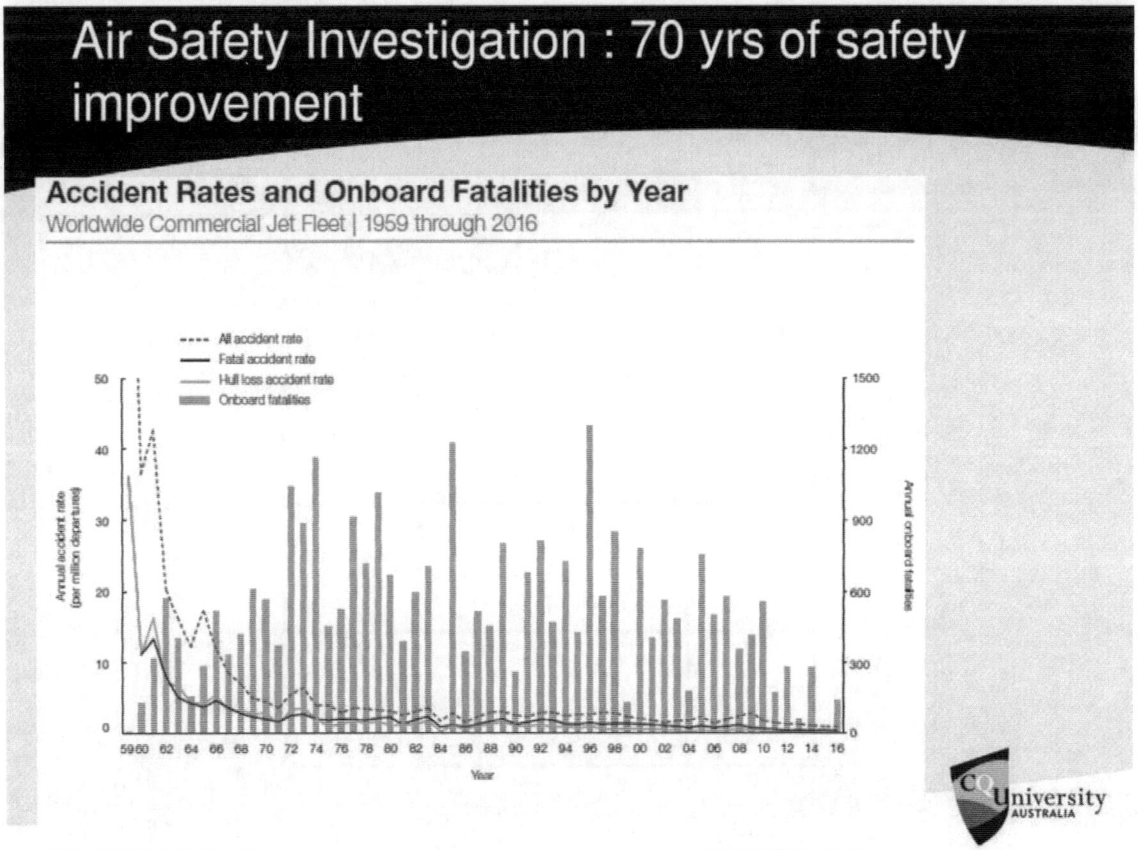
Air Safety Investigation : 70 yrs of safety improvement
Accident Rates and Onboard Fatalities by Year
Worldwide Commercial Jet Fleet | 1959 through 2016
All accident rate
Fatal accident rate
Hull loss accident rate
Onboard fatalities
Annual accident rate (per million departures)
Annual onboard fatalities
Year
16 | 2016 STATISTICAL SUMMARY, JULY 2017
Copyright © 2017 Boeing. All rights reserved.
CQUniversity AUSTRALIA

Evolution of Methods 1926

- Heinrich - Accident Modelling commences

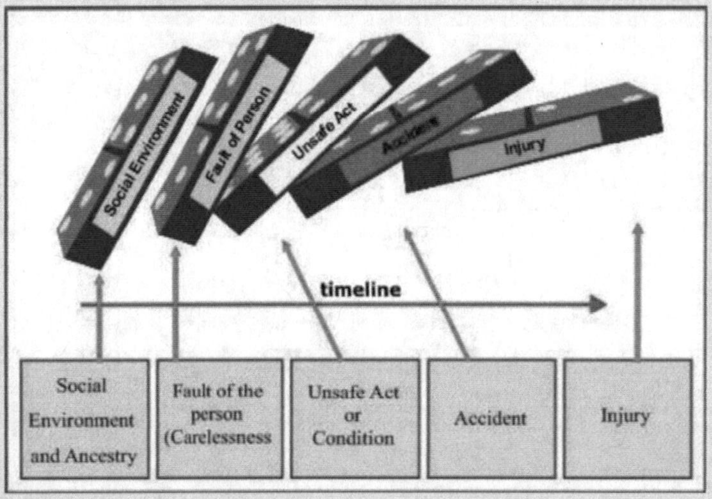

CQUniversity AUSTRALIA

Heinrich H. (1941), *Industrial Accident Prevention: A Scientific Approach*, 2nd Edition, McGraw Hill, New York.

Evolution of Methods 1964

Haddon Matrix

	Host	Equipment	Environment Physical	Environment Social
Pre-Event				
Event				
Post-Event				

W. Haddon, Jr.; E. A. Suchman; and D. Klein, Accident Research, Methods and Approaches, Harper and Row, 1964.

CQUniversity AUSTRALIA

Evolution of Methods 1973

Haddon's Energy Transfer Model

	Strategies	Illustration
1	Prevent marshalling in the first place	Prevent workers from climbing to high places from which they may fall
2	Reduce the amount of energy which is marshalled from which accidents may result	Reduce the number of workers permitted to climb to high places
3	Prevent the release of energy which has built up.	Build guard rails to prevent falls
4	Slow down the release of energy.	Reduce the height at which employees are permitted to work; slow down the rate at which explosives are permitted to burn
5	Separate, in space or time, the energy which is released from the object susceptible to injury	Prohibit entry to blasting areas during blasting periods.

Haddon W., (1973), Energy Damage and the 10 countermeasure strategies, Journal of Trauma, 13(4), 321-331

Evolution of Methods 1973

Haddon's Energy Transfer Model

	Strategies	Illustration
6	Place a physical barrier between energy source and the object susceptible to injury.	Require workers to use safety helmets, shoes or goggles
7	Modify the contact surface by rounding or softening the edges	Design cars with padded dashes; build toys without sharp edges
8	Strengthen the object against damage by energy release	Require fireproof building construction. Require workers to be vaccinated against disease
9	Mitigate the damage that has not been prevented by previous eight measures	Use fire alarm systems, sprinkler systems, emergency medical care facilities, storm-warning systems
10	Use rehabilitation and restorative techniques where damage has occurred	Retrain injured workmen with permanent disabilities

Haddon W., (1973), Energy Damage and the 10 countermeasure strategies, Journal of Trauma, 13(4), 321-331

Evolution of Methods 1979

- Extended Energy Damage Model

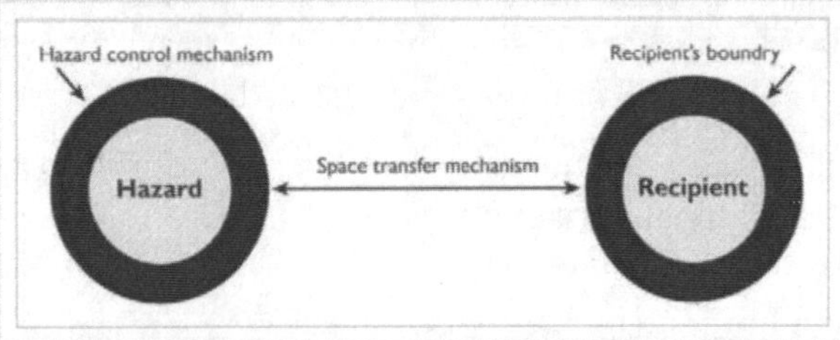

- Generalised Time Sequence Model

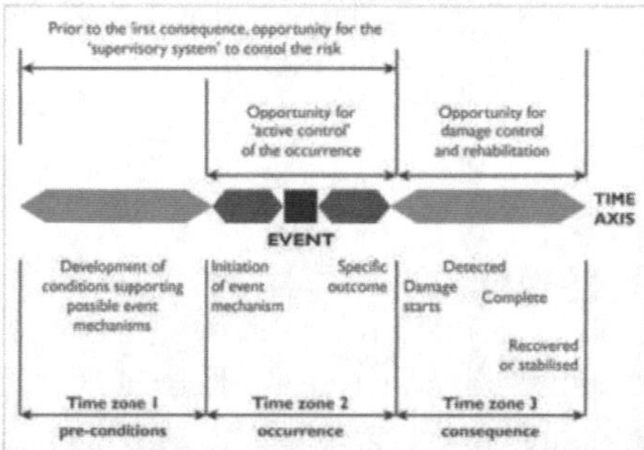

Viner, D. (1991), *Accident analysis and risk control*, Derek Viner Pty Ltd, Melbourne.

Evolution of Methods 1997

- Systems models
  - Reason

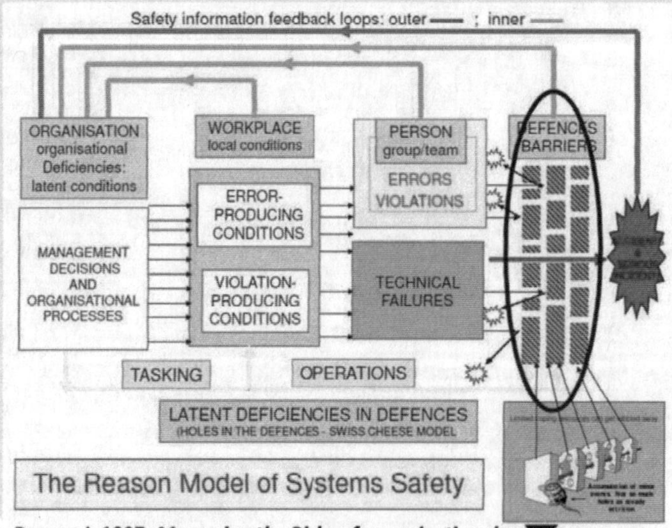

Reason J, 1997, *Managing the Risks of organisational Accidents*, Ashgate Publishing, Aldershot

Latent failure and organisational accidents

18th November 1988

DHC6 Twin Otter VH-AQB Dunk Island

Evolution of Methods : 2000

- HFACS

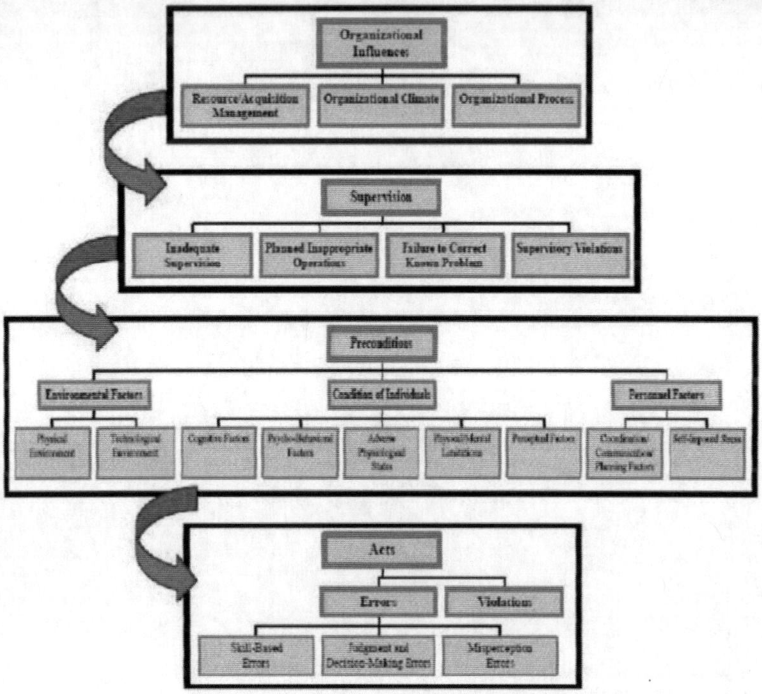

Evolution of methods : 2004

- STAMP

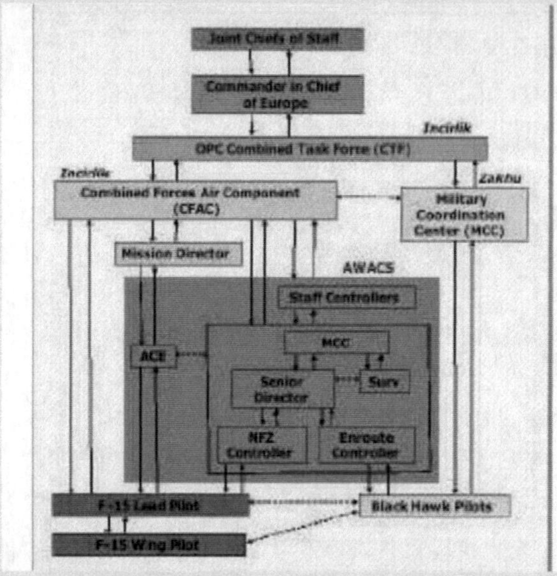

STAMP Hierarchical Command & Control Structure of the Black Hawk fratricide

(Qureshi, 2007)

Apted, R 2012 Overview of Systemic Modeling Approaches
https://www.slideshare.net/stargate1280/overview-of-systemic-modeling-approaches

2012 Hollnagel's Functional Resonance Analysis Model (FRAM)

2012

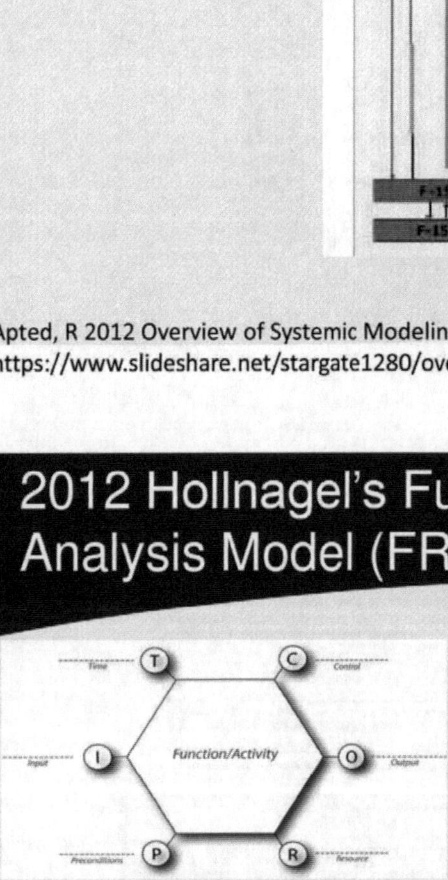

Hollnagel E, 2012, FRAM: *The Functional Resonance Analysis Method: Modelling Complex Socio-technical Systems*, Ashgate, Farnham

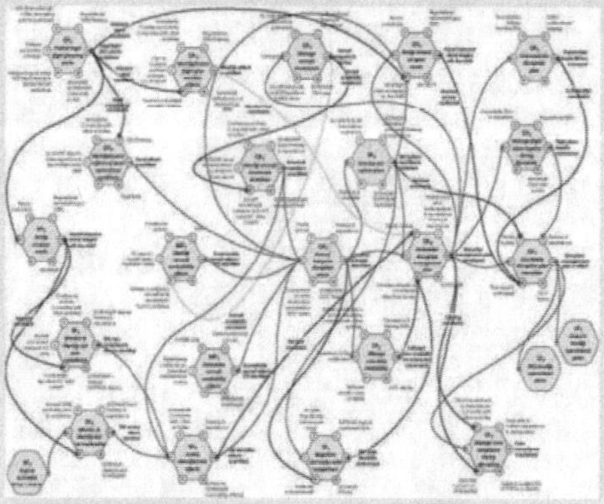

Frost B. (2014) System Hazard Analysis of a Complex Socio-Technical System: The Functional Resonance Analysis Method in Hazard Identification,
https://www.semanticscholar.org/paper/System-Hazard-Analysis-of-a-Complex-Socio-Technical-Frost/08a2470c948a4bfaff7986031e470389e7a4287c

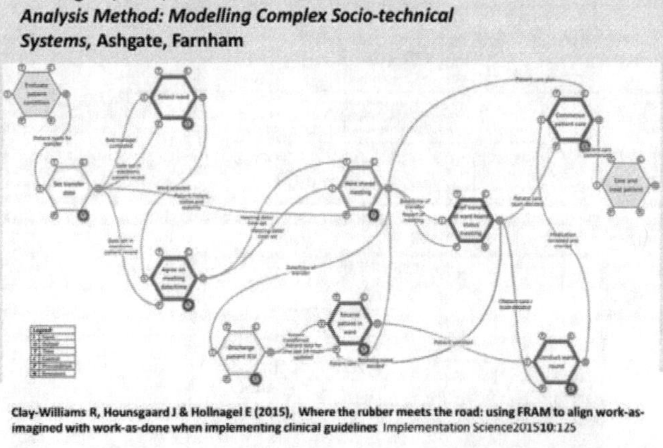

Clay-Williams R, Hounsgaard J & Hollnagel E (2015), Where the rubber meets the road: using FRAM to align work-as-imagined with work-as-done when implementing clinical guidelines Implementation Science201510:125

Evolution of Methods

- ICAM
- PEEPO

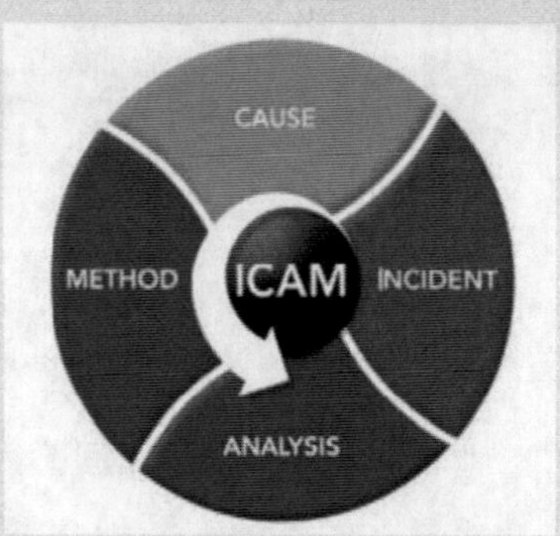

Courtesy SafetyWise Solutions

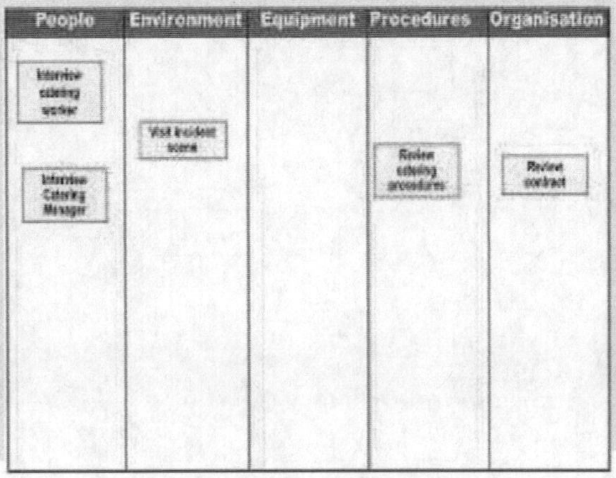

Courtesy SafetyWise Solutions

CQUniversity AUSTRALIA

Limitations

- Very few of these models provide any guidance on how an investigation ought to be conducted
- Very nearly all provide a theoretical framework for analysing evidence and data once collected
- The training needed within organisations that adopt these models needs to ensure investigators understand methods for collection, and fidelity of evidence and causation data

CQUniversity AUSTRALIA

Accident Investigation Techniques
I Know How to Do This – Don't I?

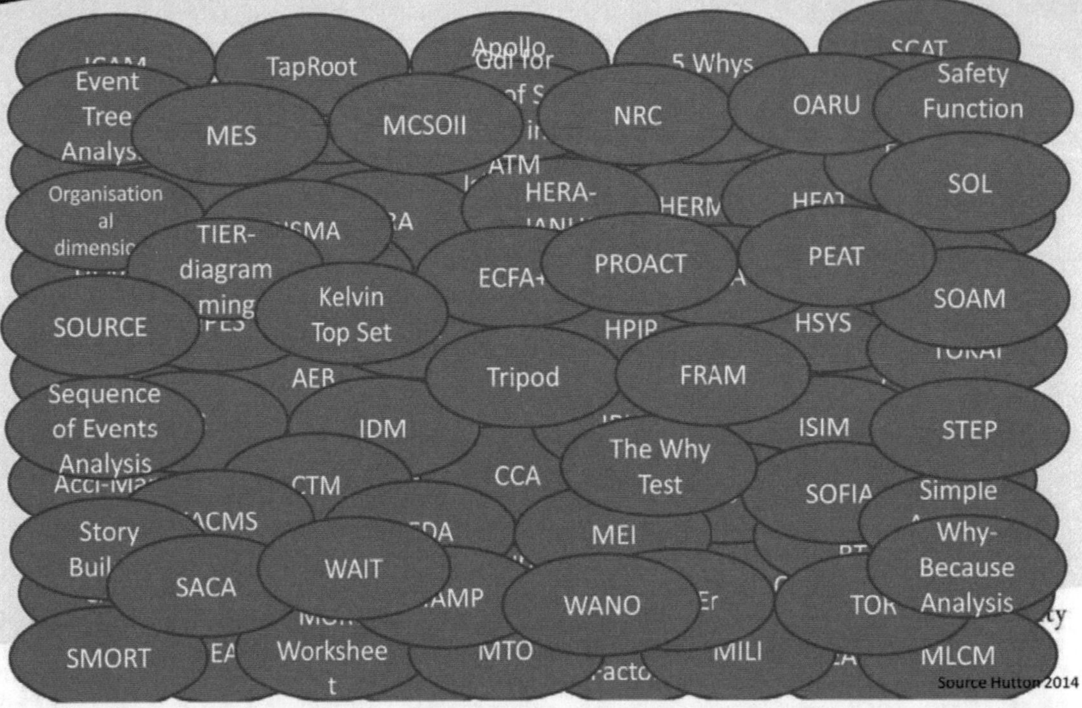

Safety II

"Don't look at what went wrong, look at what went right"

CQU's Adjunct Prof.
Erik Hollnagel

CQUniversity AUSTRALIA

Artificial Constructs

- All accident models are artificial constructs intended to help us simplify and make sense of the ever-increasing complexity of the socio-technical systems involved in accidents
- When the models appear to be as complex as the circumstance they're trying to explain, the value of the model is questionable

CQUniversity AUSTRALIA

Context is everything

CQUniversity AUSTRALIA

Context is everything

- Many models and methods provide little investigation guidance to users
  - Often users are provided with a limited number of headings to drive thinking and analysis
  - Most categorisation models do not provide any guidance on establishing causation sequences
  - Often gaps in causation understanding eventuate
- Opportunities for investigator bias and pet theories emerge unchallenged
- Shoehorning by Inexperienced Investigators:
  - Trying to fit real world evidence into the model instead of the model aiding understanding of the real world
  - Taking preconceived investigation conclusions and force them to fit the investigation model
- Some investigations lack logic and reasoning in the establishment of causal factors included in the models

CQUniversity AUSTRALIA

Logic Diagrams (Fault Trees and Event Trees)

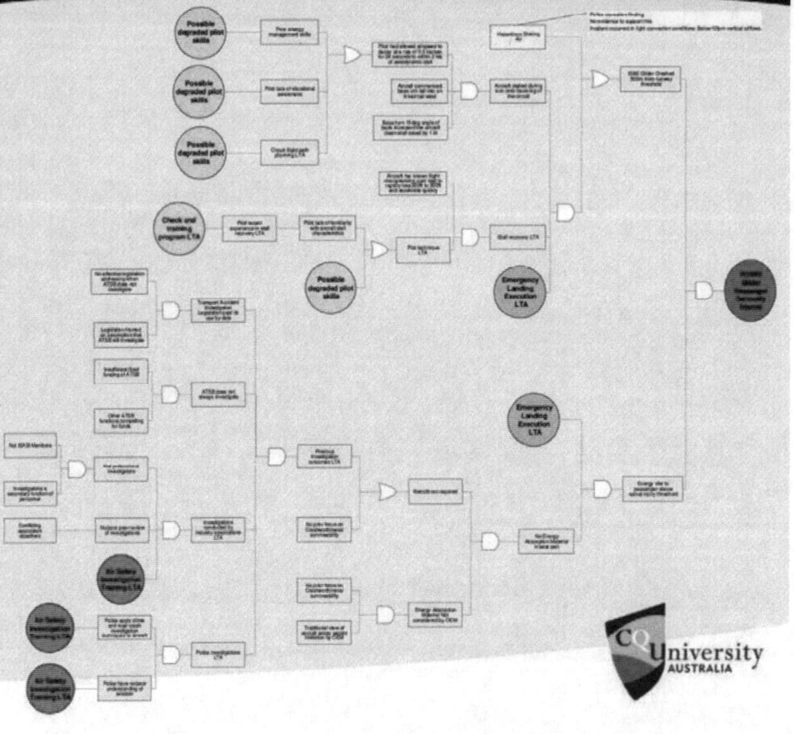

CQUniversity AUSTRALIA

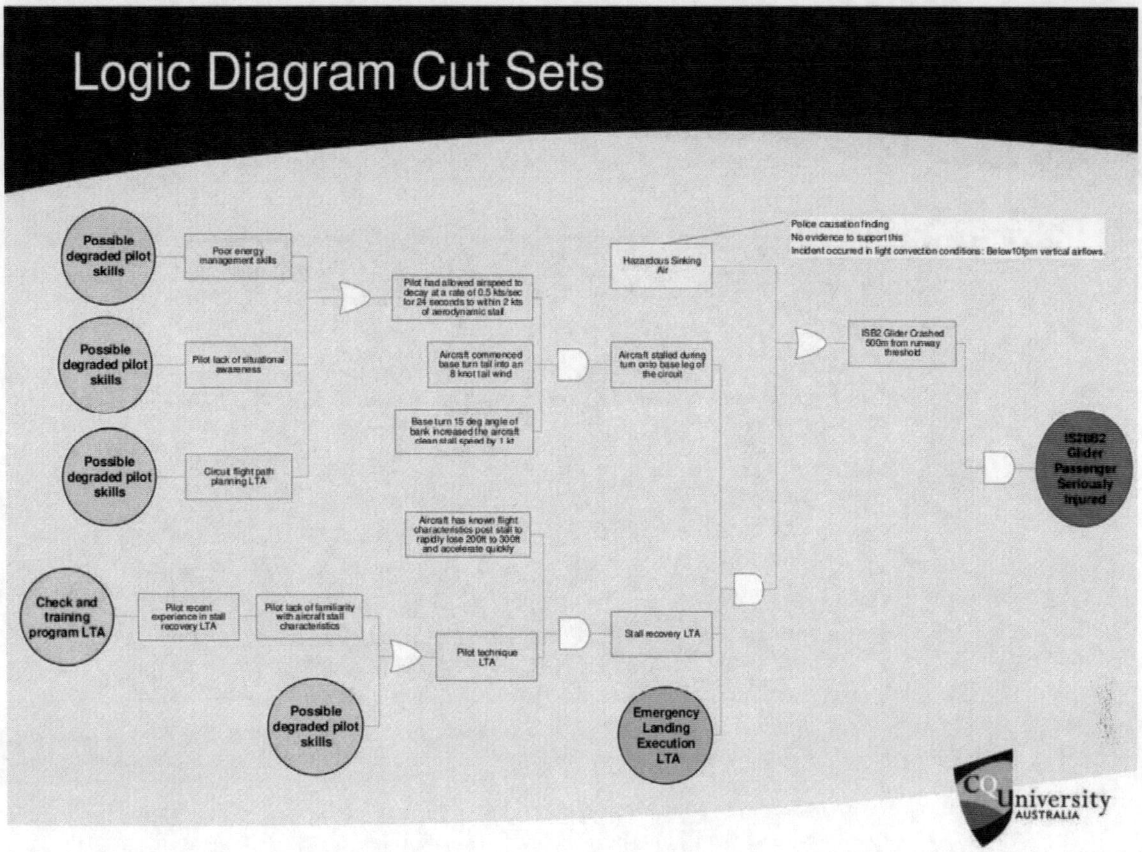

Logic Diagram Cut Sets
Possible degraded pilot skills
Poor energy management skills
Possible degraded pilot skills
Pilot lack of situational awareness
Possible degraded pilot skills
Circuit flight path planning LTA
Pilot had allowed airspeed to decay at a rate of 0.5 kts/sec for 24 seconds to within 2 kts of aerodynamic stall
Aircraft commenced base turn tail into an 8 knot tail wind
Base turn 15 deg angle of bank increased the aircraft clean stall speed by 1 kt
Hazardous Sinking Air
Police causation finding
No evidence to support this
Incident occurred in light convection conditions: Below10fpm vertical airflows.
Aircraft stalled during turn onto base leg of the circuit
IS28B2 Glider Crashed 500m from runway threshold
IS28B2 Glider Passenger Seriously Injured
Aircraft has known flight characteristics post stall to rapidly lose 200ft to 300ft and accelerate quickly
Check and training program LTA
Pilot recent experience in stall recovery LTA
Pilot lack of familiarity with aircraft stall characteristics
Pilot technique LTA
Stall recovery LTA
Possible degraded pilot skills
Emergency Landing Execution LTA
CQUniversity AUSTRALIA

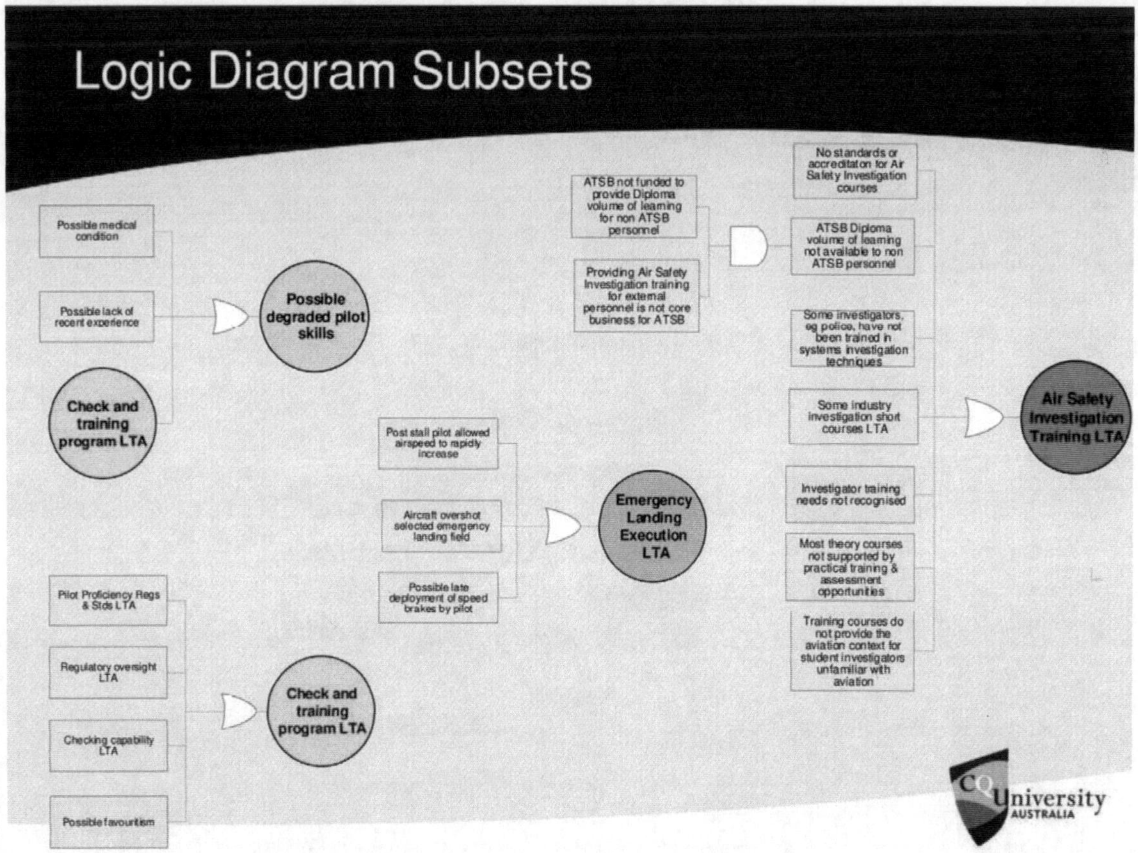

Logic Diagram Subsets
Possible medical condition
Possible lack of recent experience
Check and training program LTA
Possible degraded pilot skills
ATSB not funded to provide Diploma volume of learning for non ATSB personnel
Providing Air Safety Investigation training for external personnel is not core business for ATSB
No standards or accreditation for Air Safety Investigation courses
ATSB Diploma volume of learning not available to non ATSB personnel
Some investigators, eg police, have not been trained in systems investigation techniques
Some industry investigation short courses LTA
Air Safety Investigation Training LTA
Investigator training needs not recognised
Most theory courses not supported by practical training & assessment opportunities
Training courses do not provide the aviation context for student investigators unfamiliar with aviation
Post stall pilot allowed airspeed to rapidly increase
Aircraft overshot selected emergency landing field
Possible late deployment of speed brakes by pilot
Emergency Landing Execution LTA
Pilot Proficiency Regs & Stds LTA
Regulatory oversight LTA
Checking capability LTA
Possible favouritism
Check and training program LTA
CQUniversity AUSTRALIA

Defending your investigation logic

CQUniversity AUSTRALIA

Evolution of Methods : Now

- SafeNet

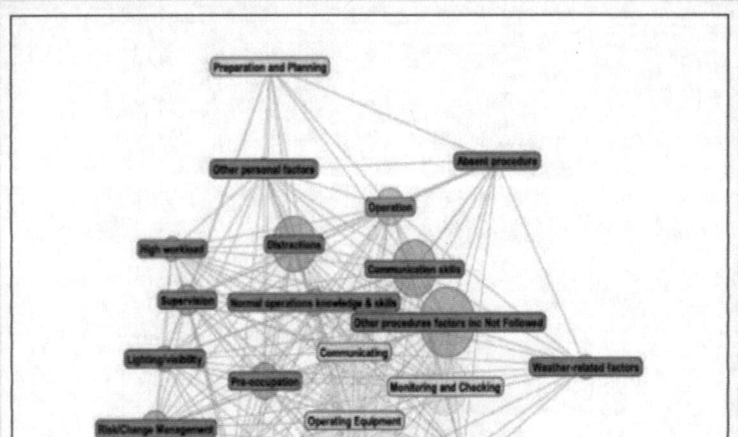

CQUniversity AUSTRALIA

Klockner, K & Toft, Y 2015, Accident modelling of railway safety occurrences: The Safety and Failure Event Network (SAFE-Net) Method Karen, 6th International Conference on Applied Human Factors and Ergonomics (AHFE 2015) and the Affiliated Conferences, AHFE 2015 , Science Direct pp 1734-1741

Where is the equivalent history of Cabin Safety Incidents?

ICAO developed the *Manual on the Investigation of Cabin Safety Aspects in Accidents and Incidents* (Doc 10062)

- Accident investigation and cabin safety improvements
- Recommended qualifications and competencies for cabin investigators
- Investigation of cabin safety and survival factors in accidents, including evacuations, turbulence encounters, fire/smoke/fumes, and decompressions
- *Incident investigations at the air operator level, including inadvertent slide deployments, unruly passengers, and in-flight medical emergencies*

How well have these concepts been applied to Cabin Safety Investigation?

Where are the Cabin Safety:

- Investigation data collection methods?
- Interviewing techniques?
- Scene photography and mapping?
- Applied logic and reasoning?
- Establishment of evidence and truth –triangulation of evidence
- Data analysis & testing methodologies
- Establishing findings and conclusions

Pretext has been to educate others about the Cabin,
not to educate Cabin Safety experts in investigation methods

CQUniversity AUSTRALIA

The future: Laser Scanning

Cabin Safety applications?
Virtual reality cabins?

Some challenges to come: Tesla Uber

Are we forgetting the lessons of the past?

Some challenges to come: Virtual Reality

CQUniversity AUSTRALIA

Some challenges to come: Drones

Are we forgetting the lessons of the past?

CQUniversity AUSTRALIA

Drones : Predator Control Station

Are we forgetting the lessons of the past?

CQUniversity AUSTRALIA

Commercialisation of the Space Industry

There will be a role for air safety investigators in this newly commercialised industry

CQUniversity AUSTRALIA

Improved investigation education for Cabin Crew Safety Professionals – a clear need

Photo coutesy of Etihad Airways

CQUniversity AUSTRALIA

WHAT'S THE DIFFERENCE BETWEEN A SAFETY INVESTIGATOR AND A JOURNALIST:

36

WHAT'S THE DIFFERENCE BETWEEN A
SAFETY INVESTIGATOR AND A JOURNALIST:

SAFETY INVESTIGATORS ARE:

SYSTEMATIC AND ACCURATE!!

Thank you

37

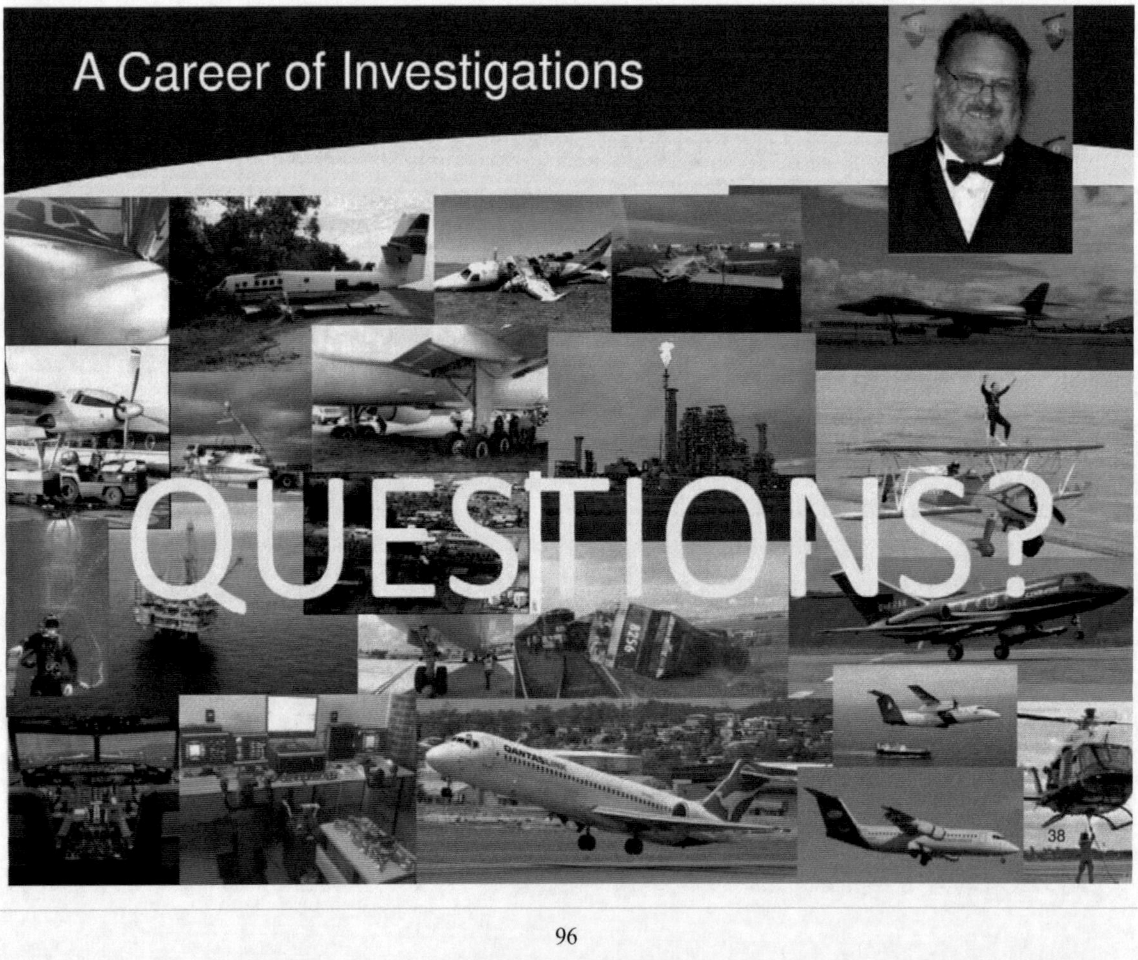

apats2018 SINGAPORE

Requisite Imagination:

Breaking the "Fiddler On the Roof" in Cabin Crew Professionalism

Joey Anca
Manager Human Risk
Metro Trains Melbourne
Australia

Joey Anca

- Manager Human Risk in Safety Critical Domains
  - Metro Trains Melbourne, Melbourne, Australia
  - Jetstar Airways, Melbourne Australia
  - Sydney Trains, Sydney Australia
  - Singapore Airlines
  - ICAO
  - Ansett Australia Airlines
  - Philippine Airlines
- Accident Investigator
  - Multiple Fatalities, Tail Strikes, SPADS, Human-Machine Scary Incidents
  - Still convincing organisations that the Just Culture must be divorced from Investigations =)!
- The CRM Book
  - Kanki, Helmreich, Anca & Chidester

apats2018 SINGAPORE

Imagine.
Why are these Word Pictures DIFFERENT?

We train our teams to meet a **Standard**

We train our teams, to demonstrate **Competency**

How do we Train our teams to become **Professionals?**

Take a Minute to Chat…

apats2018 SINGAPORE

Imagine.
Why are these Word Pictures DIFFERENT?

We train our teams to meet a **Standard**

We train our teams, to demonstrate **Competency**

How do we Train our teams to become **Professionals?**

apats2018 SINGAPORE

Requisite Imagination

(adapted from Westrum, R. (1993). Cultures with requisite imagination. Eastern Michigan University: Ypsilanti, MI. USA)

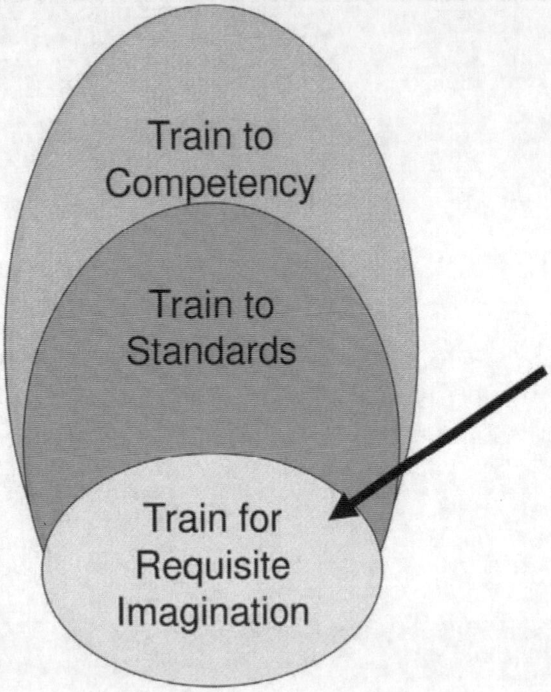

The Fine Art of Anticipating What Can Go Wrong

apats2018 SINGAPORE

$1M=1 Fatality

Halfway to
Hell Club

Maestros and Disasters

apats2018
SINGAPORE

apats2018
SINGAPORE

Professionalism: The Individual View

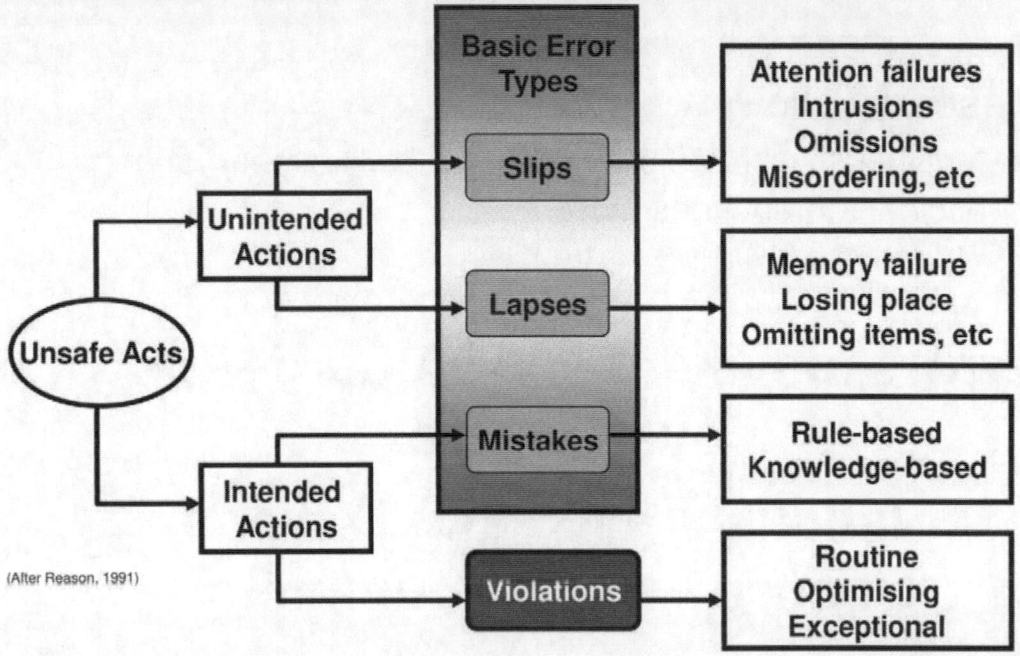

(After Reason, 1991)

apats2018 SINGAPORE

Professionalism: the Normal Operations View

(from Maurino, D (2003). Normal Operations Monitoring. ICAO: Montreal)

Errors & Violations

- Research shows:
  - Individuals who Violate, increase the likelihood of an accident three times more.
    (Helmreich, 2002)

Errors are like the VISA Card—it's everywhere

Violations are like our Driver's Licence—know your limits

apats2018 SINGAPORE

The "Fiddler On the Roof": Making Professionalism Happen

(Anca, J. (2018). APATs 2018. Singapore)

What we should see in an Integrated Flying Operations Human Factors Program

		ACTION Statements
Philosophy (Corporate Beliefs & Values)	**What is the Business' belief about the inevitability of Human Error?**	A corporate statement from the CEO that: **Human Error is part of business enterprise. The Business is committed to manage Error before it manages its People.**
Policy	**Does the Business have a process for dealing with Human Error? Is Automation a part of the Business' Policy?**	**•Articulate (Cabin Crew) Behaviour •Be Explicit about how 'At Risk' Behaviour is managed**
Procedures	**•Are Human Factors concepts part of my Operating Procedures/Manuals? •How are errors Investigated & Managed?**	HF Principles are integrated into: **•Rostering Practices/ D&A/ Fatigue Management & Accident Investigation •Training (Recurrent and Remedial) •Company Risk Management**
Practice	**How does the Business collect information about Human Error?**	**•Non-jeopardy observation/collection of error management data •Human Factors Initial and Reinforcement Training •Crew Resource Management (CRM) •Safety and Human Factors are dominant part of KPIs**

apats2018 SINGAPORE

Professional Sleeping For Cabin Crew

APATS2018 speech by James Nixon

(PowerPoint slides are in **BOLD**)

"I was the Purser, walking with the crew to the aircraft and we were stopped in the aerobridge for a random drug test. The Captain and I were tested, and we both passed. No alcohol or cocaine, but I was off my face on Zanax, a legal Benzodiazepine.

Ultra Long Haul flights killed me. I was a walking zombie. I had been OK flying short haul, but when I started flying across the Pacific, I couldn't handle it. I was going to have to resign.

Then I was given one Zanax tablet, and I slept for 13 hours straight. Just what I needed. Just what the Doctor ordered.

Soon I was Doctor-shopping, on a four year downhill spiral. In the end I could only sleep three hours at a time, and was taking Zanax just to "feel normal".

I detoxed in hospital for 8 days ... and have been clean for 123."

Those were the words of a Cabin Crew member, who, I must add; does not work for my old airline.

27 million Zanax prescriptions will be filled in the USA this year, making it the 23rd most-prescribed pharmaceutical medicine in that country.*

And it was announced in Australia yesterday, that deaths from Benzodiazepine overdoses over the last 15 years are averaging 472 a year.

[Slide of JN]
I am James Nixon, author of five books and a retired A380 Captain.

I flew for over 31 years, had about the same number of hours as Sully when he went into the Hudson River, and Captain Shah from MH370.

I flew Boeing 727s, Airbus A320s, 321s, 330s, 340s and 380s and have worked in Australia, Malta, England, Vietnam, Bahrain and Dubai. Besides flying, I studied Management, and Acting.

You can't make a career out of flying, be it as a Pilot or Cabin Crew, without getting to grips with sleep.

 james@CrammondMedia.com

Quality sleep, when you want, for as long as you want. I call it: Professional Sleeping.

Since Accountants, MBAs and Risk Managers have taken over the running of airlines, there a whole bunch of experts in middle management, sitting around in head office, giving themselves awards for generating the most 'productive rosters' each month. Getting Cabin Crew to work harder than ever before.

Because the airline industry is so young, and because they are yet to learn from history —a glance at the Mining Industry would reveal that these problems have been solved by smart people a century ago— the airlines are classic silo businesses.

Recruitment stands alone, trying to put names in the spaces on the spreadsheets being produced by Cabin Crew Training, who are filling the requirements of Cabin Crew Operations; while HR exits those crew who can't survive the pace.

The Cabin Crew churn-rate in many airlines has increased, as many give flying away as a bad joke; leave and have to be replaced by new people.

Unlike our generation, who fell in love with the smell of jet fuel as kids, and would fly for free if required; Gen Y —the Millennials— youngest 24; and Gen Z —the i-Gen— oldest 23; aren't in love with flight. For them they have seen no change. A computer is a computer, a phone is a phone, all planes - and cars - look the same.

In my day cabin crew had a gap-year before starting university, took-up flying to get away from their family, or an enthusiastic boyfriend, and made it a career, maybe retiring to start a family.

These days they choose a gap year after university, before entering their chosen career, and come to the airline to see the world. Why stay at a back-packers when you can stay at the Sheraton in New York? In my time at Emirates I had cabin crew who were qualified Doctors, Scientists, Accountants and Lawyers from 130 nationalities.

So that side of the business hasn't changed in recent times. Neither has the cabin crew from a poor background who sees this industry as an escape from the farm. They will stay forever.

So why the increase in the churn rate? Why are they leaving?

"I am so tired! This job is killing me!"

I got so sick of spending flights telling my FOs and Cabin Crew how to sleep properly, I wrote a book about it.

One retired A380 Captain wrote a review:

> *"I gave up a career that I loved, in large part because the night shifts, and subsequent fatigue, were killing me! Had this book been available, I might still be flying"*

What did it cost our airline, to replace him?

As trainers, it's your responsibility to teach them how to do all of this job. Not just safety and security. Not just first aid and cabin service. Not just responsible service of alcohol and presentation. But all of the job. And that includes sleep.

There are twenty factors that affect your sleep, to a greater or lesser degree:

Alcohol.
Stop after two glasses or you'll sleep like a dead person for 90 minutes, then spend the rest of the night bouncing off the walls.

Bed Bugs.
Check your Room, Never put your suitcase on the bed, use the stand, keep suitcase closed at all times, put dirty washing in sealed plastic bags.

Body Clock.
Learn your 90 minute sleep cycle, observe daily body cycles, plan for jet-lag. Make your body work for you - not against you.

Coffee.
Takes 15-30 mins to work then stops you sleeping for 6 hrs.
Use it strategically, before you need it. It's a diuretic, so you need to drink an equivalent amount of water.

Illegal Drugs.
Despite the fact that taking the wrong pill could cause you to sleep forever —they usually stop you sleeping properly.

To expand your mind, use a book, podcast or audiobook instead. And just say No.

My whistle-blowing Zanax user told me that one cabin crew told him that he couldn't do an Ultra Long Haul flight *"without Tina."*

My mate said: *"I thought he meant he liked listening to Tina Turner in crew rest. It turns out that Tina is code for crystal methamphetamine."*

There's a new campaign that just started in the U.S.: "One Pill Can Kill"

Legal Drugs.
Sleeping pills and Benzodiazepines are used after major traumas. None are an option for aircrew whose career lasts more than two weeks.

They should never be used with alcohol and are banned in some countries.

Like that fantastic restaurant near the hotel in Osaka, once Cabin Crew find something they like, they go there every trip.

We are addicted to things that make our lives better, so the chances of stopping your sleeping pills after two weeks —as recommended by their manufacturers— are zero.

And everyone drinks alcohol with them, which is a no no.

Just don't start.

Fitness.
Simple exercises and stretches. 30 mins brisk walking a day, every day, forever. Keeps depression at bay and helps you sleep better.

iPods, Audiobooks & Apps.
One of my theories on life is: '*If you can't win - CHEAT*' and cheat you must, if you are to make sleeping your number priority in life. Audiobooks, Podcasts, Sleeping Apps and Sleep Music. All these can be used to ensure you are ready to sleep when you want.

Massages.
Before and after ultra long-haul flights, after any long day really, a relaxing massage can prepare you for sleep like nothing else.

Happy beginning …

Mosquitoes.
Get educated, Get serious, Get safe. 216 million cases of malaria occurred in 2016, resulting in 445,000 deaths. Not to mention the other mosquito-transmitted diseases: Dengue Fever, Nile Fever, Encephalitis and Chickungunya.

Despite airlines training their crews, and providing Malerone for free, against Malaria, there is an undercurrent lie being spread in cabin crew ranks that Malerone is more dangerous than malaria.

This is not true. I can only find one case of liver damage in the world caused by Malerone, and the jury is still out on the cause. Like the flu vaccine, Malerone is upgraded each year, and the current version is OK to take almost continuously.

One thing is for sure.

If you are trying to sleep in a hotel room in Accra, Ghana; and there is one, fat, female, pregnant, Anopheles mosquito in the dark with you … sleep will be very hard to find.

Natural Remedies.
Are all fakes: Melatonin and Valerian. They don't work for Cabin Crew in a moving sleep cycle roster, and only cost you money. The best result is that they will fool you into believing they work.

Nicotine.
Affects your sleep. The smoking habit, which is harder to break than heroin in many people, even wakes you up, to be satisfied. Yet, stopping is easier than you think.

Rosters & Diaries.
Are the key to getting control of your life. Give up the electronic diaries and get back to paper. Plan sleep like you plan your work. Spend a couple of hours planning your month when you get your roster. Within three months you won't believe how great you feel.

Snoring is a sign you need to change your life:
- lose weight
- stop smoking
- reduce drinking
- cut out tranquillisers
- take up exercise or,
- go back in time and change your parents!

Tackling snoring leads to better sleep.

Sleep Apnea, on the other hand:
Shortens your life. The good news is that you won't out-spend your retirement savings.

Get medical help, and don't give up on the CPAP machines.

Stress.
Do a simple stress test. Take action, learn the tips & tricks, to reduce stress. Get help. You'll sleep like a ***baby***.

Wetting the bed every few hours and then wake up screaming … but you know what I mean.

Too Much Sun.
Cabin crew from western countries, and especially colder ones, think that lying beside the pool in the tropics, without sunscreen, is clever.

It isn't.

Sunburn prevents restful sleep, then skin cancer kills you.

Too Little Sun results in long-term sleep issues, and people from Arctic countries have to take special measures to be able to sleep in summer, when they sun never sets, and in winter, when it never rises.

They call it the SADs, Seasonal Affective Disorder, a type of depression associated with the seasons. Get help, to be able to sleep better.

Water.
Well-hydrated humans sleep better. It also reduces cellulite and makes your skin look better. While cabin crew complain about working in a dry environment, few drink enough water, often saying that they are worried about having to go to the toilet during crew rest breaks.

The thing is, your wondrous human body will adapt after a couple of weeks.

Weight & Sex.
Getting control of weight issues leads to better sleep, and sex induces sleep, *so they tell me!*

That's the twenty factors that affect your sleeping.

But it's how you trick your subconscious, and the tips required to get to sleep, that we have to also teach our crews.

Prepare for sleep. **[Sleep Preparation]**

I am going to reset your sleep clock. Here comes a yawn. And many of you are going to yawn too.

Y A W N

And in one and a half hours, at 11:45, you are going yawn again.

Then again at 1:15.

And again at 2:45.

And later, at 4:15.

And then, at 5:45. We are surfing the 90 minute sleep cycle wave.

At ten past eleven something great is gonna happen. You are gonna be in bed. Beside you is a pad and pen, and you have written down the things in your mind. Hotel dressing gown blocking-out the crack under the door. Pitch black room, three alarms set, earplugs in, sleep mask on your forehead, phone on airplane mode. TV off. Room slightly cool, you will be snuggled-down reading a book by one solitary light. It doesn't matter what the book is. The light is reflecting off the pages.

Suddenly the sleep wave comes, and you are about to yawn.

Book down, light-off, eye-shade on. Sleep.

By planning your sleep before a flight, using tips and tricks, you know you are going to sleep naturally for, say six and a half hours. You block off eight hours, and wake naturally.

With three alarms set, your subconscious can relax, it knows that you are going to wake up and doesn't have to keep bothering you all night … *'You awake? Remember, you have to go to work soon.'* It does that all night.

Three alarms, the last one is gonna wake up the whole neighbourhood. It knows that you are going to wake up, no matter what, so can let you sleep deeply.

By using these tricks you'll wake up naturally, all month, and never hear an alarm. Except for those flights where I want you to wake up after three hours for strategic reasons. Maybe two or three times a month.

You need to get your **Room, Bed, and Sleep-Position** right.

To use **Self-Hypnosis & Relaxation Techniques**. Even Pre-Sleep Affirmations, with the goal of **waking-up naturally**.

Learn how to create a great state of mind, at work. To **fake it til you make it**. So when someone asks you how you are … say fantastic!

You soon will be.

Someone who makes each flight zoom along; energising your fellow crew-members, not dragging them down by complaining about your roster.

You need to master sleeping in the **Crew Rest [Coffin]** Compartment. Not lying there, wide awake, in a coffin.

And use the **NASA Nap**. 24 minutes that will change your life. Forever.

Discover that the **Layover Destination** starts on board the aircraft. Take control of your personal security, learn the truth about elevators, and hotel fires. Answer the questions so your subconscious will let you relax.

Really understand **Jet Lag**. These days it's only really for passengers, since we're back at home before the dreaded Third Day. By becoming an expert you can help your passengers.

Learn how to survive the **Expat Lifestyle**, if you have to go overseas to work.

Surviving **A Long Flight**.

The actual experience of being in a aeroplane for a long time has little to do with sleeping, or jet lag. You can fly from Moscow to Johannesburg for 14 hours and 30 minutes in the same time zone and still feel like you've been hit by a truck. I discovered *'23 Tips To Survive A Long Flight'* but they have little to do with **Professional Sleeping**.

Professional Sleeping is all in your mind.

They say that an audience like this will only remember 3 things. They lie.

Before our much-needed coffee break, if you are anything like me, you will be lucky to remember my name. James as in James Bond. Nixon as in President. Google me. Email me and let me know if I am right.

To make a successful career in aviation and look this good when you retire after 31 years, remember these three things:

1 Sleep has to be your **#1 priority**, plan your sleep.

2 Three **alarms, earplugs and eyeshades**.

3 **[No Electricity.]** No phones. No TV. No Ipad. No Computer.

Oh, and number 4. There are 4,416 airliners up above us right now. A million passengers. And, lets say, 40,000 cabin crew. How many are taking Zanax to *'feel normal'*?

Most airline crashes happen on the first sector after an overnight.

Are your crews ready?

Have they slept, professionally?

I hope so.

[Contrail Slide]

*Footnote: *Valium (Diazepam) vs. Xanax (Alprazolam) Comparison* click HERE

***

Listen to this speech at ***The Nixon Tapes Podcast*** **HERE**. (24 mins)
[CAUTION: Language warning!]

**

James Nixon is available for speaking engagements: **HERE**.
He travels from Melbourne, Australia.

*

www.CrammondMedia.com

apats2018
SINGAPORE

SUSAN RICE EMAIL:TAASUSAN@GMAIL.COM WEDNESDAY 29 AUGUST CABIN SAFETY CONFERENCE 1

apats2018
SINGAPORE

"Challenges to education of the aviation travelling public"

Susan Rice – Aviation Consultant/Cabin Safety Specialist
Email: taasusan@gmail.com
Wednesday 29 August 2018

apats2018 SINGAPORE

Is there a genuine need for education of our passengers?

SUSAN RICE - TAASUSAN@GMAIL.COM WEDNESDAY 29 AUGUST CABIN CREW CONFERENCE

3

apats2018 SINGAPORE

SUSAN RICE EMAIL:TAASUSAN@GMAIL.COM WEDNESDAY 29 AUGUST CABIN SAFETY CONFERENCE

4

apats2018 SINGAPORE

- *How do we approach this challenge?*
- *Is it a global issue?*

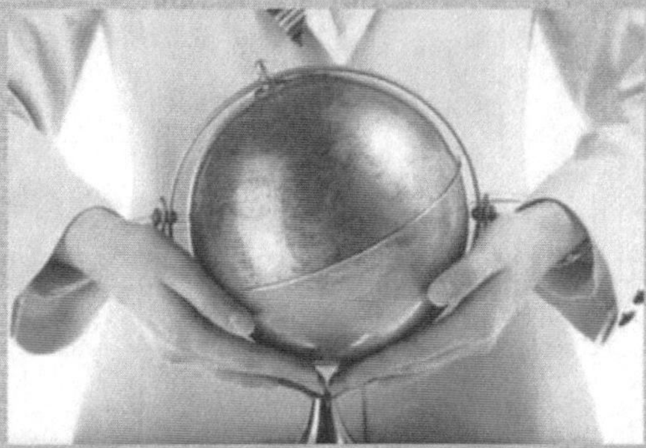

SUSAN RICE - TAASUSAN@GMAIL.COM WEDNESDAY 29 AUGUST CABIN CREW CONFERENCE 5

apats2018 SINGAPORE

Do we shock passengers?

Would images such as these achieve the desired outcome?

SUSAN RICE EMAIL:TAASUSAN@GMAIL.COM WEDNESDAY 29 AUGUST CABIN SAFETY CONFERENCE 6

apats2018 SINGAPORE

SUSAN RICE EMAIL:TAASUSAN@GMAIL.COM WEDNESDAY 29 AUGUST CABIN SAFETY CONFERENCE 7

apats2018 SINGAPORE

What is the desired outcome of this education?

SUSAN RICE - TAASUSAN@GMAIL.COM TUESDAY 29 AUGUST 2017 CABIN CREW CONFERENCE 8

apats2018 SINGAPORE

- *What do we want to impart to the travelling public that they do not already hear about and see in the movies or on television?*

SUSAN RICE EMAIL-TAASUSAN@GMAIL.COM WEDNESDAY 29 AUGUST CABIN SAFETY CONFERENCE 9

How do we begin our journey of education? apats2018 SINGAPORE

ICAO Doc 10086
Manual on Information and Instructions for Passenger Safety - First Edition 2018
Foreword

"ICAO standards address the need for passengers to receive safety information on board aircraft. Well-informed, knowledgeable passengers have a better chance of surviving a life-threatening situation on board an aircraft."

SUSAN RICE EMAIL-TAASUSAN@GMAIL.COM WEDNESDAY 29 AUGUST CABIN SAFETY CONFERENCE 10

- In Australia research has been conducted into passenger receptiveness to delivery of the safety message:
  - Australian Transport Safety Bureau - ATSB
    "Public Attitudes, Perceptions and Behaviours towards Cabin Safety Communications" Andrew Parker, Synovate Pty Ltd
    Report Number:B2004/0238, 2006

    Abstract

"Analysis identified that low levels of passenger attention to safety communications results from overconfidence, superficial familiarity with messages, issues relating to the way safety content is presented, perceptions of substitutability between the card and briefing and social norms present in the aircraft cabin"

Cont.

Executive Summary

"Of all communications types tested, the safety briefing was most prone to perception of reduced relevance through repeated exposure, while very low attention levels and perceptions of content established safety cards as being generally ineffective. Passenger enjoyment, derived from a recognised measure of media effectiveness, was found to be low for the safety demonstration. This was particularly so among males and frequent flyers, who interestingly, were also found to be the least likely to engage with safety messages and the least likely to demonstrate desirable attitudes to cabin safety communications"

apats2018 SINGAPORE

"Examining the Effectiveness of Pre-Flight Cabin Safety Announcements in Commercial Aviation"

Published in The International Journal of Aviation Psychology, 2009

Dr Brett Molesworth, Associate Professor, School of Aviation - University of New South Wales

CONCLUSION

" Many airline passengers ignore pre-flight safety announcements in the form of a video presentation. Arguably in acknowledgement of this, some airlines have altered the way they deliver this important safety announcement. The research reported here tested three variations of pre-light safety videos and found that employing a celebrity to introduce the material, or creating a pre-flight safety briefing that is topical, fun, and entertaining (i.e. humorous), are the best methods of facilitating the retention and recall of the information in the video.........."

SUSAN RICE EMAIL:TAASUSAN@GMAIL.COM WEDNESDAY 29 AUGUST CABIN SAFETY CONFERENCE 13

apats2018 SINGAPORE

"Airline Passengers' Awareness of and Preferred Source of Cabin Safety Information"

Yu-Hern Chang, Meng-Yuan Liao
Asia Pacific Management Review 15(4) (2010) 533-547, 2004

CONCLUSION

"Finally, the most important conclusion derived from the present study is that current cabin safety information presentation need to be substantially improved, because only 5 of the 26 questions indicated that our respondents knew abut Cabin Safety information at the mean level of "agree". The safety demonstrations provided by Cabin Crews have changed very little over the years. A more attractive, specific, and comprehensive cabin safety demonstration is necessary. Also, different approached of conveying cabin safety information to the public need to be considered"

SUSAN RICE - TAASUSAN@GMAIL.COM WEDNESDAY 29 AUGUST CABIN CREW CONFERENCE 14

apats2018 SINGAPORE

Increased delivery of additional specific safety information pre-flight:

- ✓ *Passengers to keep shoes on for take-off and landing*
- ✓ *Stronger emphasis on passengers to keep seat belt fastened*
- ✓ *Window shades open for take-off and landing*
- ✓ *Cabin Pressure failure: Aircraft will immediately commence a rapid descent – normal operational requirement!*
- ✓ *Specific attention to brace position*
- ✓ *Greater emphasis paid to identifying nearest exit*

apats2018 SINGAPORE

Should the delivery of education only be about the pre flight safety demonstration?

- ❑ *OBVIOUS ANSWER =* ***"NO"***

What are some options?

- ❑ *TV screens in airport terminals*
- ❑ *Airline websites*
- ❑ *Visiting schools and talking to the children*
- ❑ *Greater clarity in onboard communications*

apats2018 SINGAPORE

Challenges to education of the aviation travelling public:

- *Sales and Marketing Departments within airline operators*

- *Passengers should not be "frightened"*
- *We shouldn't "scare" our passengers*
- *The more aware a person is the better they will respond to a given situation – ICAO have been documenting this notion for decades!!*

SUSAN RICE EMAIL:TAASUSAN@GMAIL.COM WEDNESDAY 29 AUGUST CABIN SAFETY CONFERENCE

Thank you !
apats2018
SINGAPORE
SUSAN RICE
EMAIL:TAASUSAN@GMAIL.COM
WEDNESDAY 29 AUGUST
CABIN SAFETY CONFERENCE
19

Human Trafficking: Moving Beyond

AVION LLC
TRAINING

A Globally Recognized Problem

How many of you are familiar with the global the problem of human trafficking?

Has anyone ever suspected trafficking was happening on one of your flights?

Have you been trained by your airline on what to do and where to report it if it happens?

Awareness and training are different: Awareness is about recognition and training is about empowerment.

16

Airlines are seriously
starting to train.

OHCHR

United Nations Office of the High Commissioner for Human Rights

Trafficking in persons means placing or maintaining individuals in an exploitive situation for economic gain. It is different from "smuggling."

It is the Transportation or harboring of individuals through force, fraud or coercion for the purpose of exploitation.

ACT

- Recruitment
- Transportation
- Transfer
- Harbor
- Receipt of persons

\+

MEANS

- Threat
- Use of Force
- Coercion
- Abduction
- Fraud/deception
- Abuse of power or vulnerability
- Giving payments or benefits

\+

PURPOSE

Exploitation including:

- Prostitution of others
- Sexual exploitation
- Forced labor
- Slavery
- Removal of organs

= **TRAFFICKING IN PERSONS**

Cabin Crews are a Key component of global response

- Aviation is one of the primary modes of transportation used by traffickers.
- Cabin crew are front-line personnel (face-to-face).
- Are in a unique position to be able to recognize visible signs of trafficking.
- Have enough time with passengers to be able to observe questionable situations.

Actively Promoting Training

UNODC
United Nations Office on Drugs and Crime

ICAO

UNITED NATIONS
HUMAN RIGHTS
OFFICE OF THE HIGH COMMISSIONER

IATA

Emirates

AirAsia

Aer Lingus

American Airlines

skyregional

jetBlue

DELTA

Comprehensive Guidelines

Providing the framework and topics that should be included in the training – aimed specifically at cabin crew.

Specialized Training for Cabin Crews

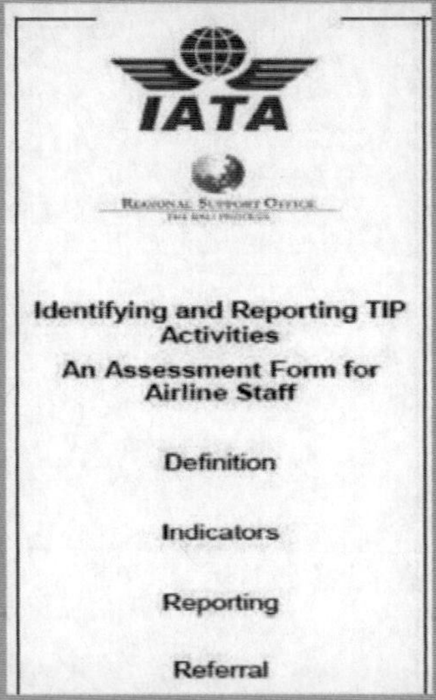

ICAO and IATA recommend each operator develop:

- A set of dedicated policies procedures and reporting protocols for when such encounters occur
- Communication protocols
- An assessment form and reporting mechanism

Areas of Emphasis

- Elements of trafficking
- Awareness/types of trafficking
- Why trafficking happens (who are victims/perpetrators?
- General indicators of trafficking
- *Indicators for cabin crew members*
- Cabin management (responding to other passenger's concerns)

Specific Recommendations

- Training should be recurrent and interactive
- delivery methods appropriate to the subject matter
- In-classroom training with videos, quiz, team exercises and concrete case studies
- Involve training other personnel
- Include reporting procedures in the flight and cabin crew manuals
- *Stress the concept of "Do no harm"*

"Do No Harm"

- Be discreet with the discussion and relaying of information so as not to raise suspicion.
- Do not confront the trafficker or rescue victim
- Act normally. Do not display unusual concern or alarm.

Going Beyond the Cabin

- Awareness campaigns on board (videos/safety cards)
- Posters
- Company's FB page
- Include in Social media campaigns

IATA

We have our #eyesopen

MOVE BEYOND AWARENESS

For a list of references on trafficking in persons email me at:
aviontraining1@gmail.com

AVION LLC TRAINING

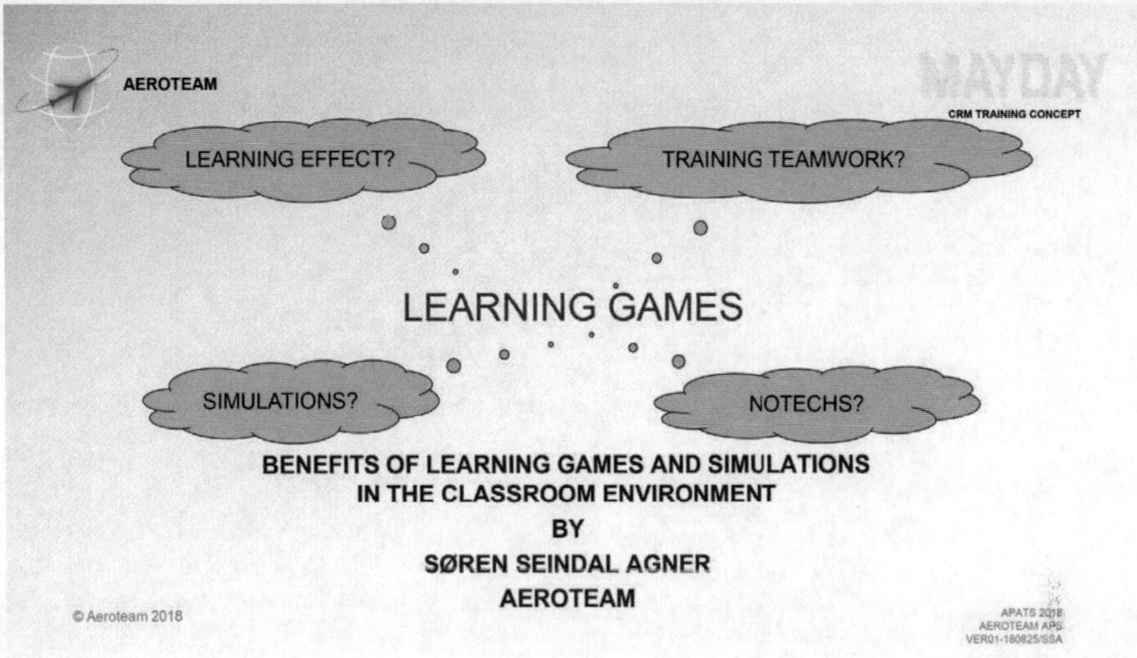
AEROTEAM
MAYDAY
CRM TRAINING CONCEPT
LEARNING EFFECT?
TRAINING TEAMWORK?
LEARNING GAMES
SIMULATIONS?
NOTECHS?
BENEFITS OF LEARNING GAMES AND SIMULATIONS
IN THE CLASSROOM ENVIRONMENT
BY
SØREN SEINDAL AGNER
AEROTEAM
© Aeroteam 2018
APATS 2018
AEROTEAM APS
VER01-180825/SSA

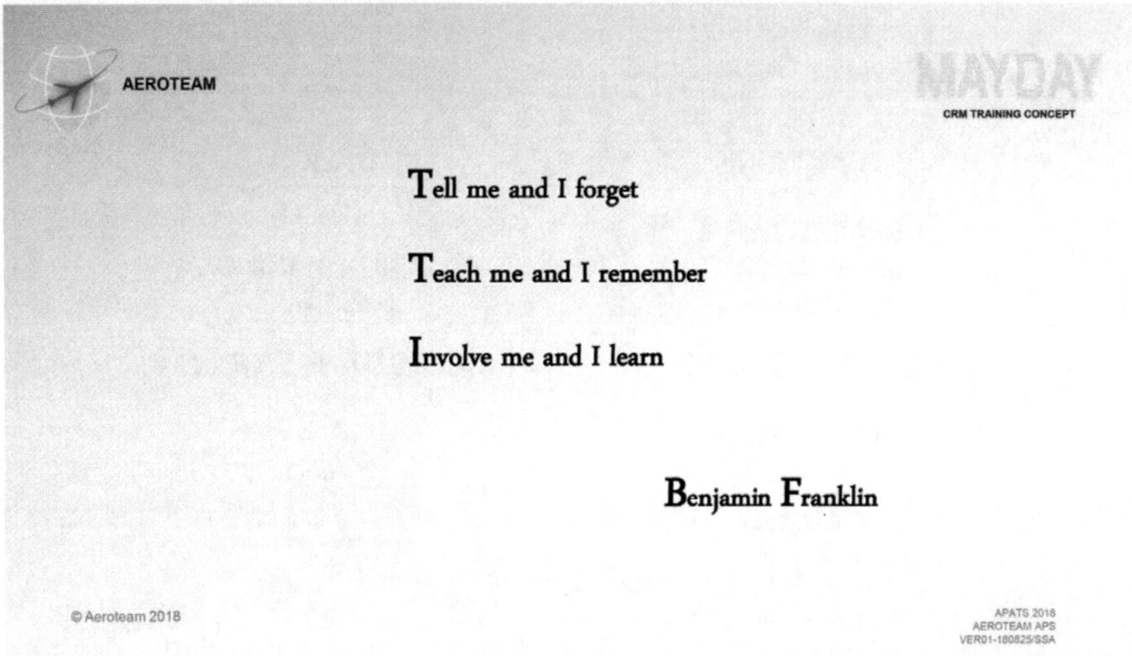
AEROTEAM
MAYDAY
CRM TRAINING CONCEPT
Tell me and I forget
Teach me and I remember
Involve me and I learn
Benjamin Franklin
© Aeroteam 2018
APATS 2018
AEROTEAM APS
VER01-180825/SSA

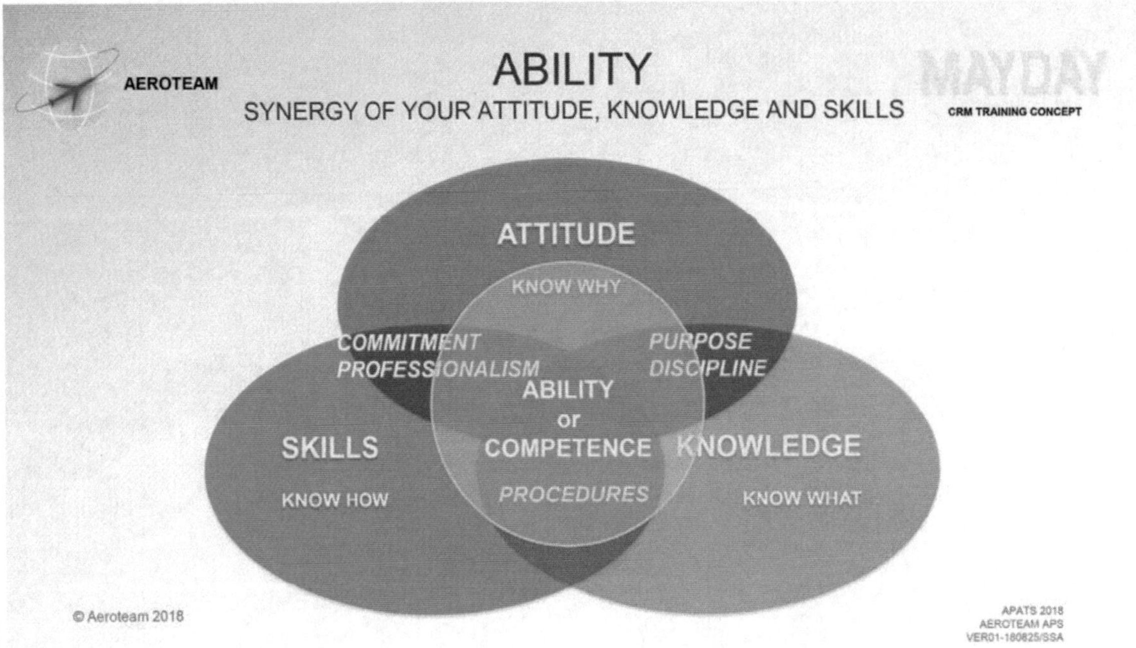
AEROTEAM
ABILITY
SYNERGY OF YOUR ATTITUDE, KNOWLEDGE AND SKILLS
MAYDAY
CRM TRAINING CONCEPT
ATTITUDE
KNOW WHY
COMMITMENT
PROFESSIONALISM
PURPOSE
DISCIPLINE
ABILITY
or
COMPETENCE
SKILLS
KNOW HOW
KNOWLEDGE
KNOW WHAT
PROCEDURES
© Aeroteam 2018
APATS 2018
AEROTEAM APS
VER01-180825/SSA

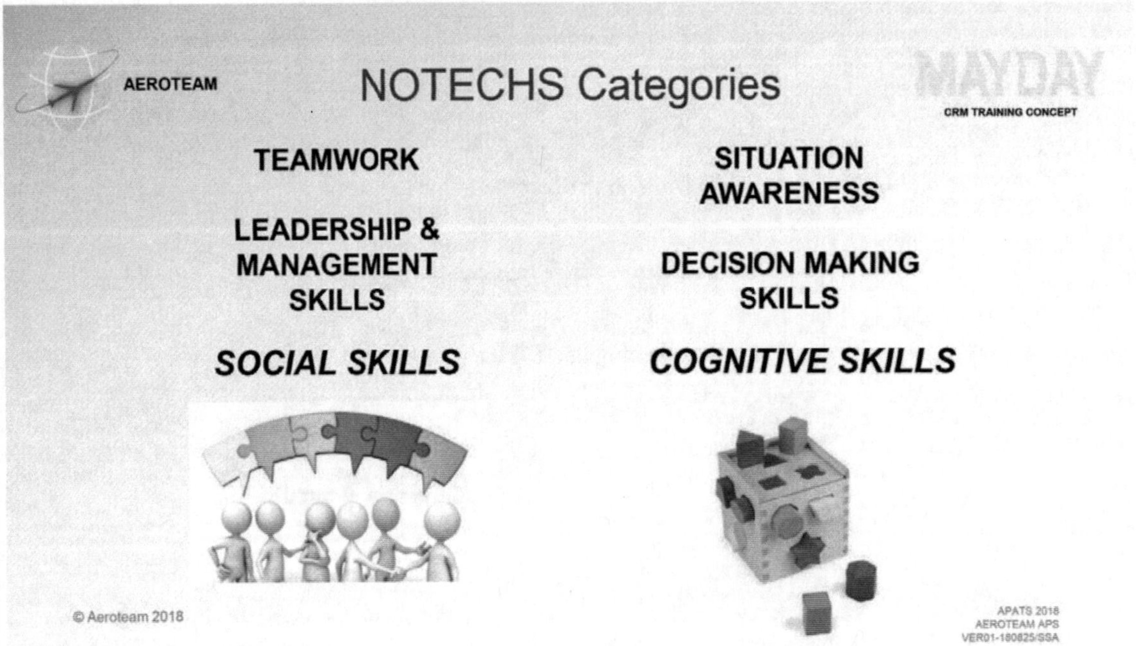
AEROTEAM
NOTECHS Categories
MAYDAY
CRM TRAINING CONCEPT
TEAMWORK
LEADERSHIP &
MANAGEMENT
SKILLS
SOCIAL SKILLS
SITUATION
AWARENESS
DECISION MAKING
SKILLS
COGNITIVE SKILLS
© Aeroteam 2018
APATS 2018
AEROTEAM APS
VER01-180825/SSA

AEROTEAM

SIMULATION

MAYDAY
CRM TRAINING CONCEPT

- Learning games are Simulations
- Learning by Doing or
  - Learning by Simulation
- Simulation creates a situation where ideas, decisions, or actions can be tried out
- Consequences are visible
- No actual risk

APATS 2018
AEROTEAM APS
VER01-180825/SSA

AEROTEAM

TRAINING

MAYDAY
CRM TRAINING CONCEPT

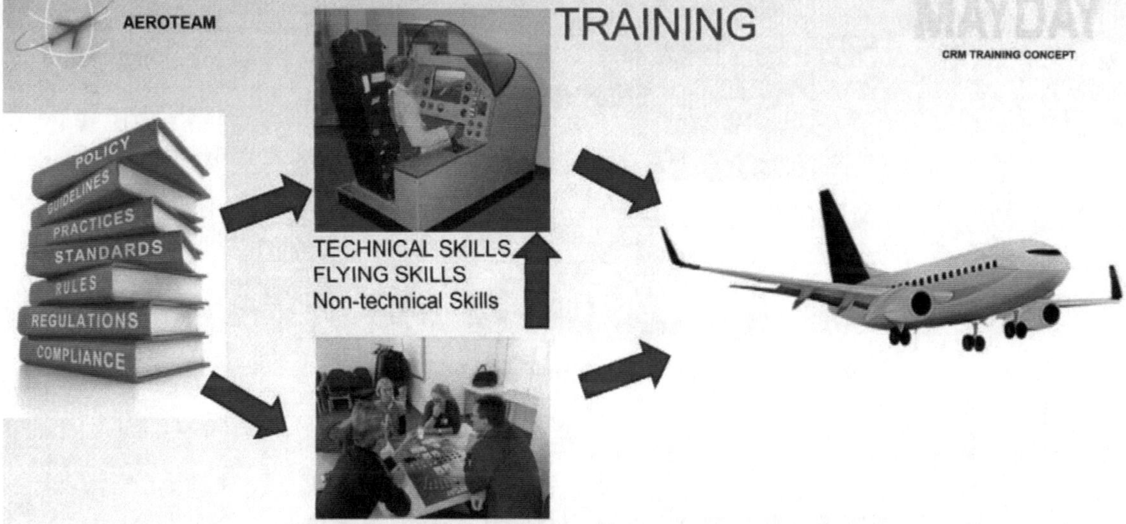

NON-TECHNICAL SKILLS
COMPETENCE and ABILITY

APATS 2018
AEROTEAM APS
VER01-180825/SSA

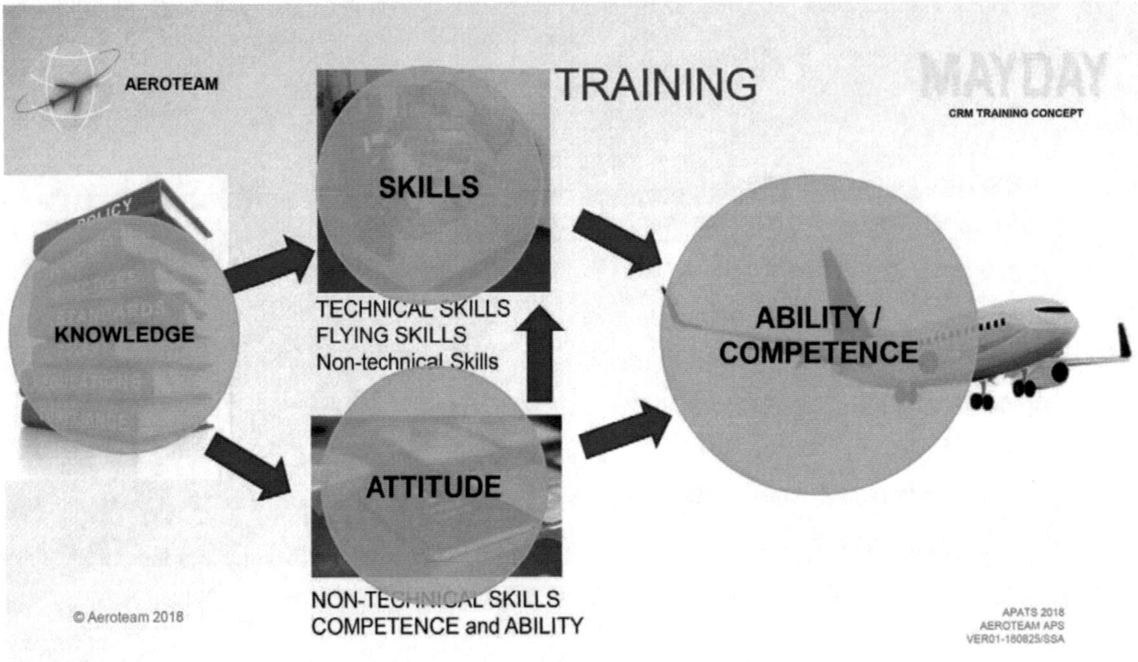
AEROTEAM
TRAINING
MAYDAY
CRM TRAINING CONCEPT
SKILLS
KNOWLEDGE
TECHNICAL SKILLS
FLYING SKILLS
Non-technical Skills
ABILITY / COMPETENCE
ATTITUDE
NON-TECHNICAL SKILLS
COMPETENCE and ABILITY
© Aeroteam 2018
APATS 2018
AEROTEAM APS
VER01-180825/SSA

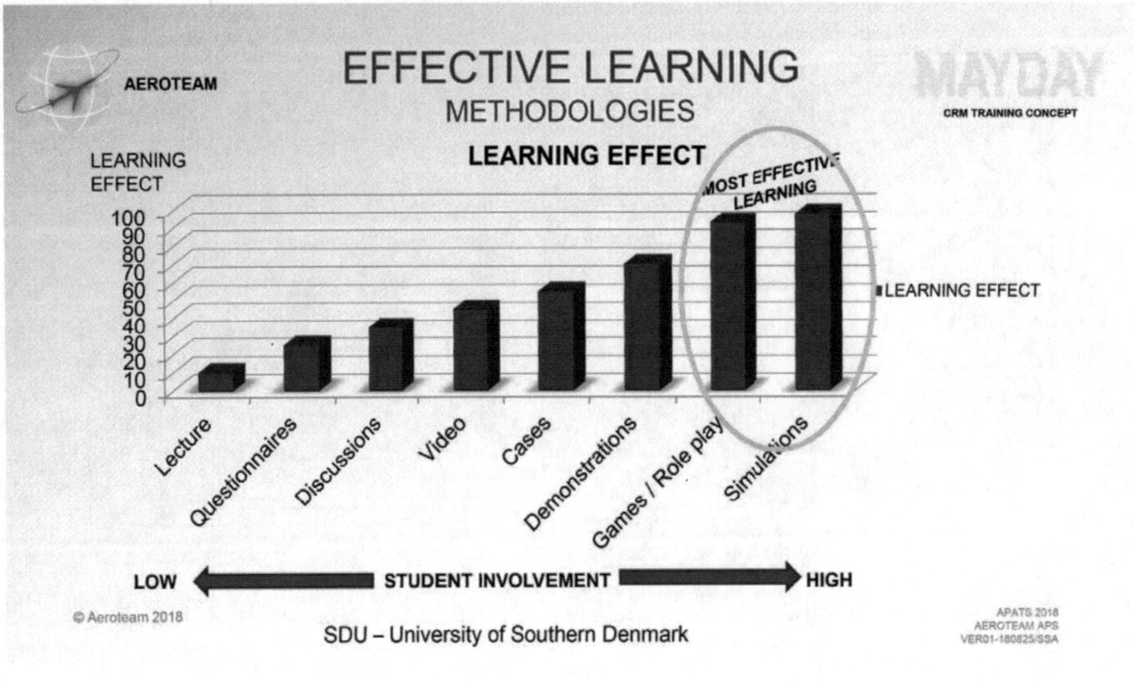
AEROTEAM
EFFECTIVE LEARNING
METHODOLOGIES
MAYDAY
CRM TRAINING CONCEPT
LEARNING EFFECT
LEARNING EFFECT
MOST EFFECTIVE LEARNING
100
90
80
70
60
50
40
30
20
10
0
LEARNING EFFECT
Lecture
Questionnaires
Discussions
Video
Cases
Demonstrations
Games / Role play
Simulations
LOW
STUDENT INVOLVEMENT
HIGH
© Aeroteam 2018
SDU – University of Southern Denmark
APATS 2018
AEROTEAM APS
VER01-180825/SSA

AEROTEAM

LEARNING GAMES

ANALOGUE OR DIGITAL

MAYDAY
CRM TRAINING CONCEPT

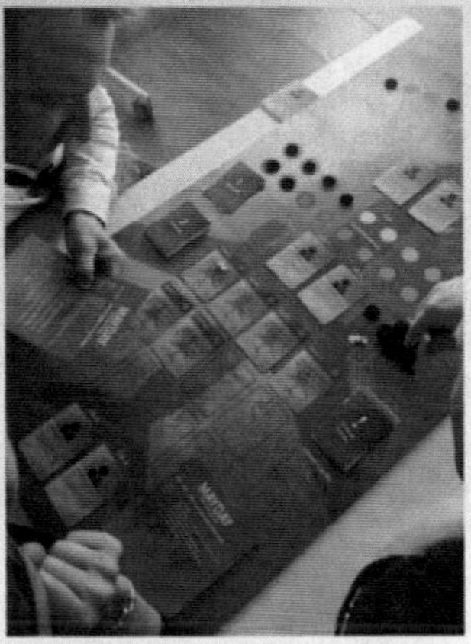

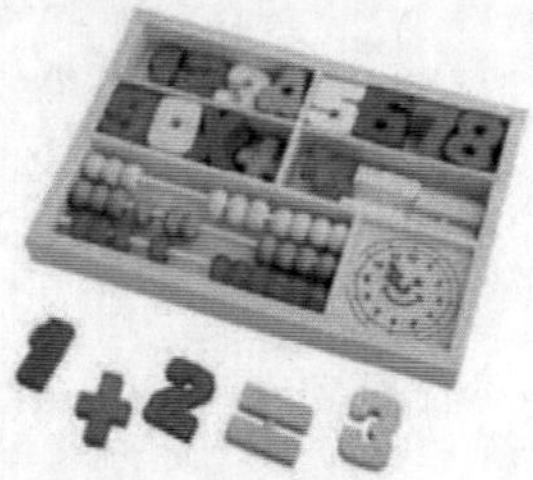

FOCUSPOINT: Learning – Not entertainment

APATS 2018
AEROTEAM APS
VER01-180825/SSA

AEROTEAM

SOCIAL LEARNING

MAYDAY
CRM TRAINING CONCEPT

- Safe to be challenged
- It is OK to make mistakes
- Mistakes can be evaluated
- Individual learning → in a dynamic social environment
- Assessment of Non-Technical Skills (observable behaviours)

APATS 2018
AEROTEAM APS
VER01-180825/SSA

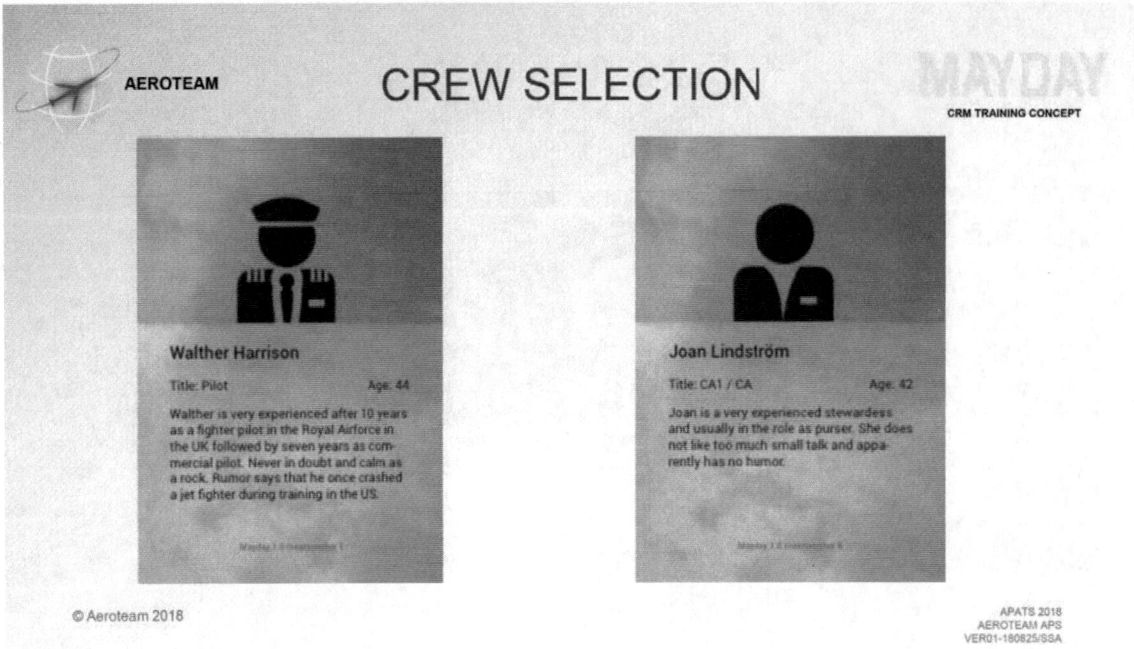
AEROTEAM
CREW SELECTION
MAYDAY
CRM TRAINING CONCEPT
Walther Harrison
Title: Pilot
Age: 44
Walther is very experienced after 10 years as a fighter pilot in the Royal Airforce in the UK followed by seven years as commercial pilot. Never in doubt and calm as a rock. Rumor says that he once crashed a jet fighter during training in the US.
Joan Lindström
Title: CA1 / CA
Age: 42
Joan is a very experienced stewardess and usually in the role as purser. She does not like too much small talk and apparently has no humor.
© Aeroteam 2018
APATS 2018
AEROTEAM APS
VER01-180825/SSA

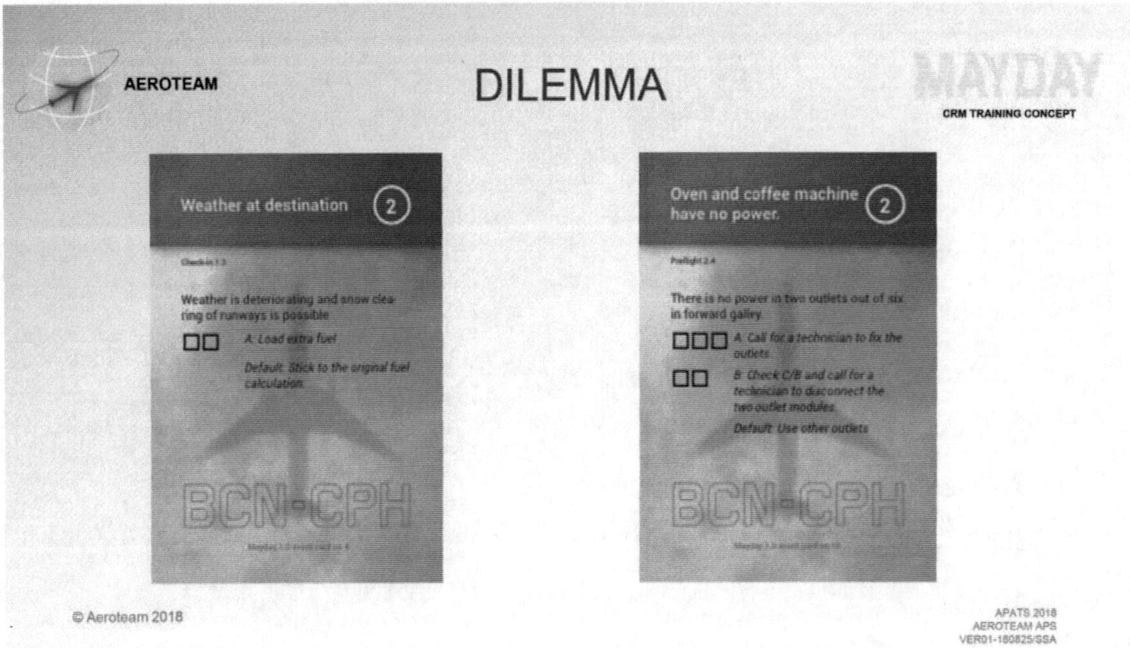
AEROTEAM
DILEMMA
MAYDAY
CRM TRAINING CONCEPT
Weather at destination
2
Weather is deteriorating and snow clearing of runways is possible
A: Load extra fuel
Default: Stick to the original fuel calculation
BCN-CPH
Oven and coffee machine have no power.
2
There is no power in two outlets out of six in forward galley
A: Call for a technician to fix the outlets
B: Check C/B and call for a technician to disconnect the two outlet modules
Default: Use other outlets
BCN-CPH
© Aeroteam 2018
APATS 2018
AEROTEAM APS
VER01-180825/SSA

AEROTEAM

9 CORE Competencies
ICAO EBT Doc 9995

MAYDAY
CRM TRAINING CONCEPT

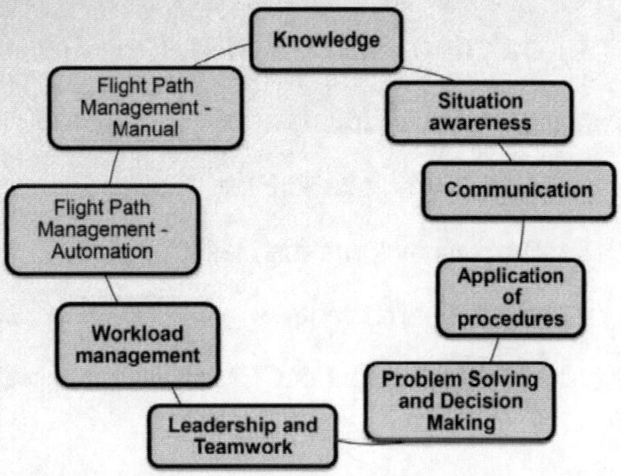

APATS 2018
AEROTEAM APS
VER01-180825/SSA

AEROTEAM

SUMMARY

MAYDAY
CRM TRAINING CONCEPT

- Introduction of Learning Games into crew training will:
  - Enhanced the learning effect by active participation
  - Enhance the Non-technical Skills of the crew
  - Connect classroom training to the real world
  - Define problem areas for future training (EBT)
  - Enhance Instructor's Competence in Assessment

APATS 2018
AEROTEAM APS
VER01-180825/SSA

AEROTEAM

SUMMARY

MAYDAY
CRM TRAINING CONCEPT

- Introduction of Learning Games into crew training will:
  - Assessment of visible behaviours
  - Increase the value of the simulator training
  - Provide cost effective, efficient and fun classroom training
  - Engage and motivate the participants
  - Provide possibility for Assessment of CRM Skills in the classroom

APATS 2018
AEROTEAM APS
VER01-180825/SSA

AEROTEAM

Summary

MAYDAY
CRM TRAINING CONCEPT

- Learning games has to be:

  Realistic – a simulation of reality

  Simple – simple presentation of complex problems

  Precise – clear and precise learning objective

  Well designed – focus on the ***learning***, not technology

  Measureable for the learning effect.

APATS 2018
AEROTEAM APS
VER01-180825/SSA

AEROTEAM

BENEFITS

MAYDAY
CRM TRAINING CONCEPT

- Benefits of Introducing Learning Games into Crew Training
  - Increase the Learning Effect of CRM Training
  - Enhancement of Simulator Training
  - Enhancement of Crew Competencies
  - Assessment of Non-Technical Skills
  - Observation of Behaviours – Future training
  - Cost Effective Training – Value for money

18
PS
VER01-180825/SSA

AEROTEAM

MAYDAY
CRM TRAINING CONCEPT

"Education is NOT the learning of facts,

It is the training of the mind to THINK!"

Albert Einstein

APATS 2018
AEROTEAM APS
VER01-180825/SSA

AEROTEAM

WHO WAS I?

MAYDAY
CRM TRAINING CONCEPT

- **Søren Seindal Agner, AEROTEAM**
  - Tour Guide, Cabin Crew, Operations
  - Pilot
    - Danish Air Transport (Short 7)
    - Scandinavian Airlines (MD80)
  - Ground Instructor – ATPL
  - CRM Trainer / Examiner
  - Multi Crew Course Instructor (MCCI)
  - Training Manager
  - Member of The Danish Flight Safety Council since 2010
    - Human Factors in Civil Aviation

APATS 2018
AEROTEAM APS
VER01-180825/SSA

AEROTEAM

THANK YOU FOR YOUR ATTENTION!

MAYDAY
CRM TRAINING CONCEPT

Aeroteam
Rørkærvej 108,
8722 Hedensted
+45 20 32 36 76
ssa@aeroteam.dk
www.aeroteam.dk

…if everything seems under control, you're not going fast enough…

APATS 2018
AEROTEAM APS
VER01-180825/SSA

Aviation Training Innovation

Over the past decade the airline training industry has pursued technology to improve efficiencies and enhance training

- Shift training earlier in the training continuum
- Allow students to better learn at their own pace, with options to use simulation

Regulatory authorities have been willing to explore use of such technology, techniques and capabilities

Although there has been substantial innovation, opportunities exist

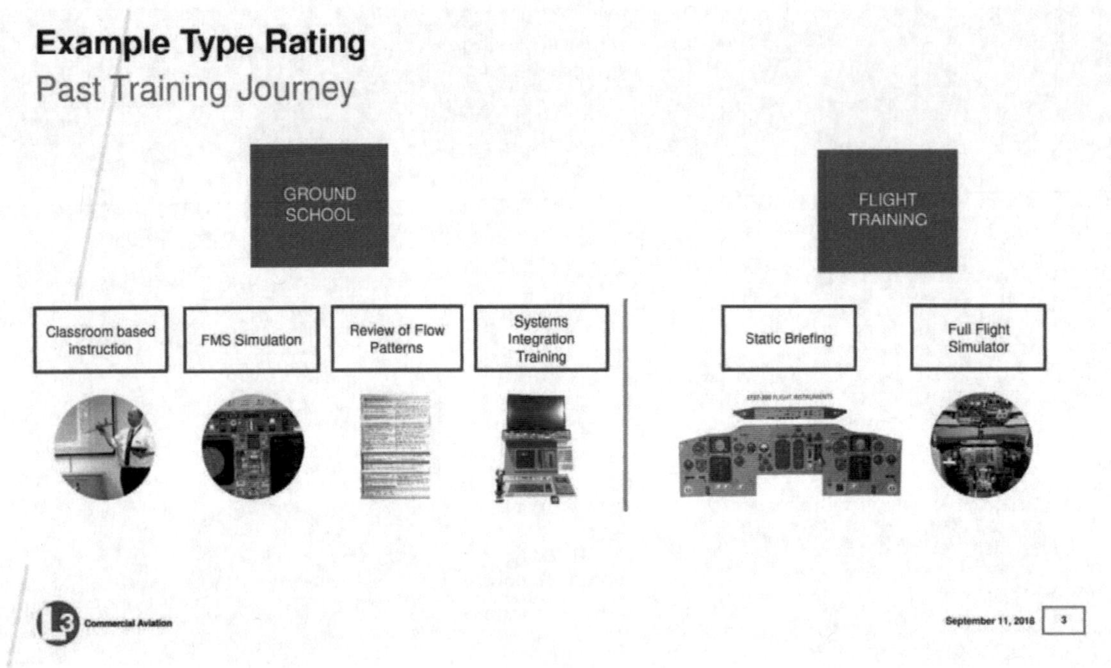
Example Type Rating
Past Training Journey
GROUND SCHOOL
FLIGHT TRAINING
Classroom based instruction
FMS Simulation
Review of Flow Patterns
Systems Integration Training
Static Briefing
Full Flight Simulator
L3 Commercial Aviation
September 11, 2018
3

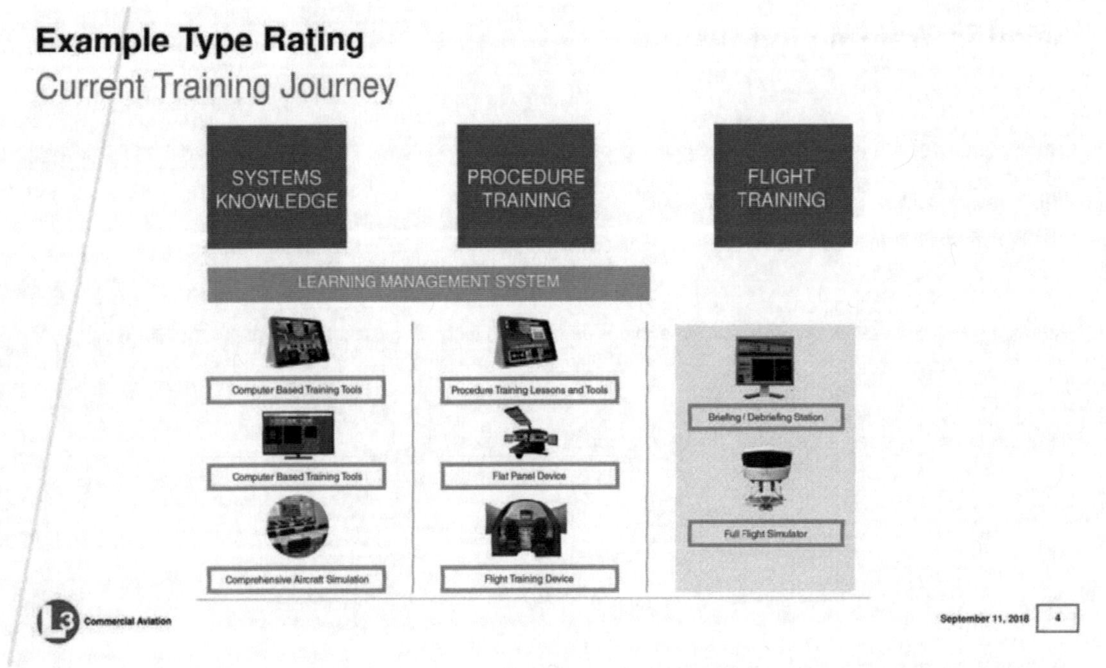
Example Type Rating
Current Training Journey
SYSTEMS KNOWLEDGE
PROCEDURE TRAINING
FLIGHT TRAINING
LEARNING MANAGEMENT SYSTEM
Computer Based Training Tools
Computer Based Training Tools
Comprehensive Aircraft Simulation
Procedure Training Lessons and Tools
Flat Panel Device
Flight Training Device
Briefing / Debriefing Station
Full Flight Simulator
L3 Commercial Aviation
September 11, 2018
4

Total Training Solution Type Rating Program

Utilizing L3 Products

SYSTEMS KNOWLEDGE	PROCEDURE TRAINING	MANEUVERS VALIDATION	LINE ORIENTED SCENARIOS	LOE / CHECKRIDE
Mobile Applications	Flat Panel Devices	FFS	FFS	FFS

FAA Type Rating

Synopsis Using Advanced Training Simulation And Technology

Although there has been substantial innovation, opportunities exist...

Traditional Type Rating Curriculum

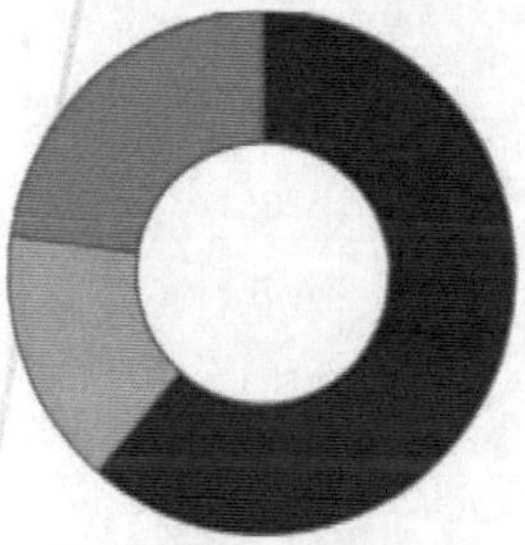

- Ground Instructor Guided Training
- Flight Simulator (Non-Motion)
- Full Flight Simulator (Motion)

Current Type Rating Curriculum

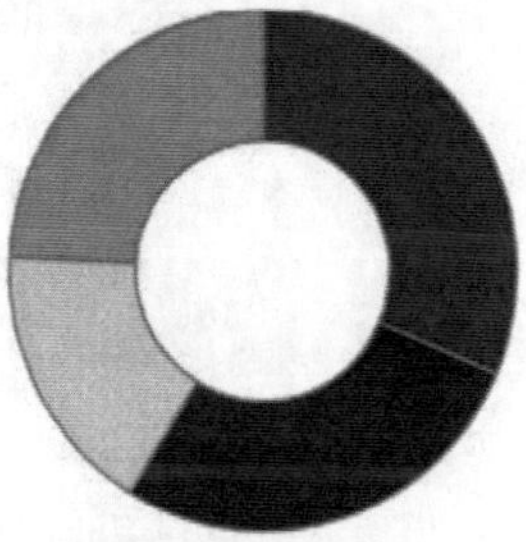

- Ground Instructor Guided Training
- Computer Based Training
- Lower Level Device
- Full Flight Simulator (Motion)

Future Type Rating Curriculum

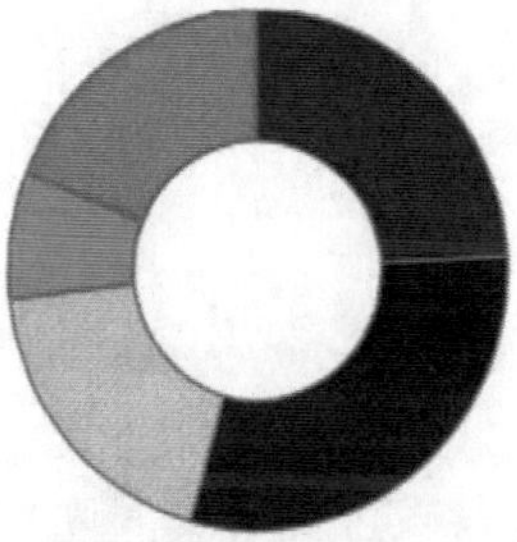

- Ground Instructor Guided Training
- Computer Based Training
- Lower Level Device
- FTD FAA Level 7
- Full Flight Simulator (Motion)

Areas Of Innovation

2
PROCEDURE TRAINING

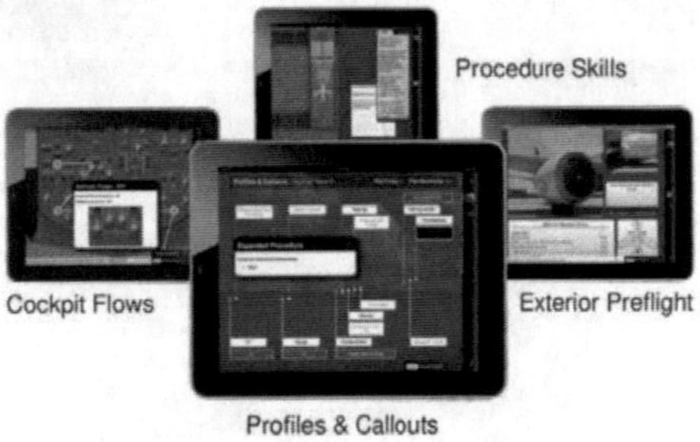

Use of Distance Learning tools to enable early learning of procedures

- Leverage technology earlier in training
- Flows, standard operating procedures, flight profile activities and callouts
- Provide training in a way 'familiar' with the next generation

Commercial Aviation

September 11, 2018 | 7

Areas Of Innovation

2
PROCEDURE TRAINING

Augmented Reality/Virtual Reality

Commercial Aviation

September 11, 2018 | 8

Areas Of Innovation

2
PROCEDURE TRAINING

Leverage Distance Learning Content on
Lower Level Training Devices

Content Guided Lessons

- Build upon the procedures knowledge gained through distance learning
- Incorporate Evidence Based Training and Competency Based Training concepts
- Empower students to obtain more training, and access to review content if desired

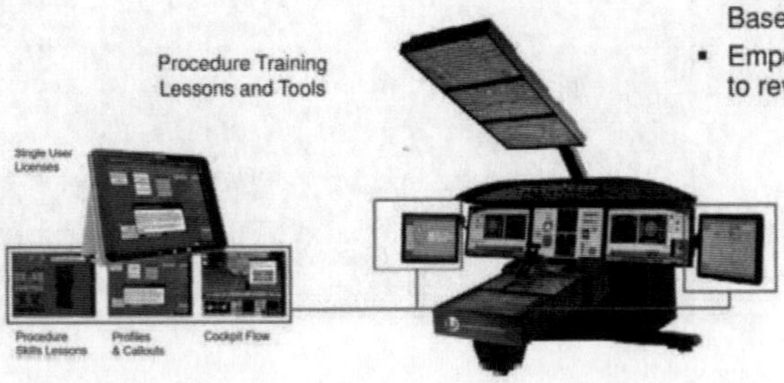

Commercial Aviation

September 11, 2018 9

Areas Of Innovation

2
PROCEDURE TRAINING

Leverage Distance Learning Content on
Lower Level Training Devices

Content Guided Lessons

- Integrate within a Training Management System
    - Track and record student actions for instructor review and analysis
- Improve Ground Instructor efficiency

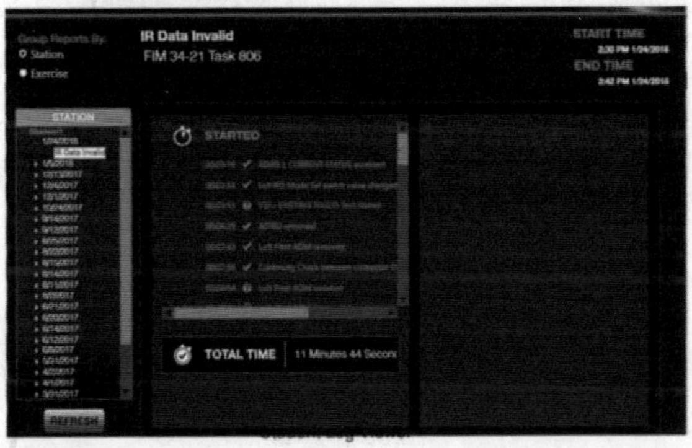

Commercial Aviation

September 11, 2018 10

Another Step Forward

World's First Level 7 FAA Device

FAA Level 7 increased fidelity and functionality including:

- FFS comparable visual scenes
- Reduced latency or transport delay – 100ms for instrument systems and 120ms for visual
- FFS software load
- Validation for First Officer controls
- Improved aural cues matching that of FFS
- Increased control feel dynamics to replicate aircraft

Fidelity meets the FFS reduced infrastructure requirements, reducing initial and operating costs compared to a FFS

Areas Of Innovation

FAA Level 7 FTD

Part 60 outlines Tasks vs FTD Level – Subjective Requirements

- Increased subjective requirements for FAA FTD Level 7
- FAA will be incorporating additional Part 60 defined FSTD levels into the pilot training and checking requirements
- Assuming the FAA updates similar to Subjective Requirements, these will have a strong impact on FTD utilization for initial training

Utilization as part of recurrent training

- Field Study for use in recurrent training (US AQP Airline)

Areas Of Innovation

FAA Level 7 FTD

Potential additional task credit for a Level 7 FTD include (majority qualified for introductory initial or recurrent qualification training):

- Taxiing
- Takeoff (normal, crosswind, instrument, engine failure)
- Windshear recovery
- Precision Approach (One engine inoperative)
- Circling Approach
- Missed Approach (One engine inoperative)
- Landings and Approach to Landing

Additional tasks acceptable for Level 7 FTD and lower (not currently noted for training credit):

- Takeoff and Departure Phase – Rejected Takeoff
- Engine Failure (Inflight Maneuver)
- Circling Approach
- Missed Approach (Normal)
- Flight Control Systems (Normal / Abnormal Procedures)

Just-In-Time

Training and operations innovation

Just-in-Time Training Capabilities

- Maintenance and Pilot Departments
- Ensure minimum level of competency during training events
- Relevance to the operations environment as well
- Why not provide tools to prepare and/or augment a crew for unique operational requirements?
  - Interactive Computer Based Training
  - Simulations
  - Videos
  - Augmented Reality

Cost Benefits
Summary

Traditional Type Rating Curriculum

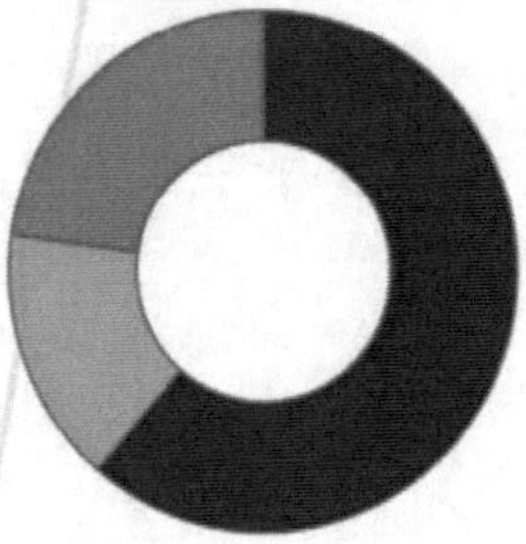

- Ground Instructor Guided Training
- Flight Simulator (Non-Motion)
- Full Flight Simulator (Motion)

With the improved training efficiencies, cost reductions can be recognized:

- Improved utilization of instructors and devices
- Efficiencies in reporting and assessment
- Reduction of training center footprint due to increased distance learning and small infrastructure requirements

Future Type Rating Curriculum

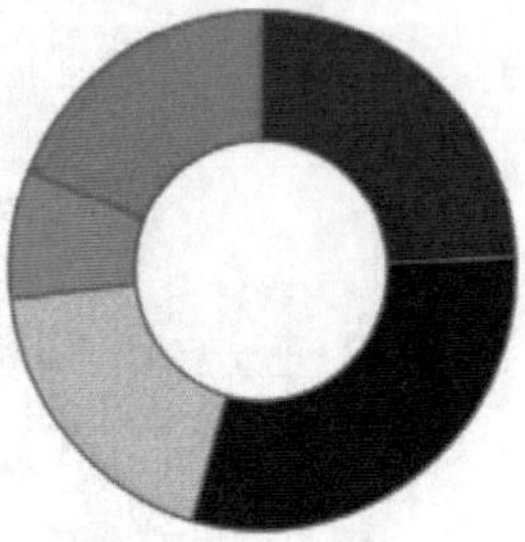

- Ground Instructor Guided Training
- Computer Based Training
- Lower Level Device
- FTD FAA Level 7
- Full Flight Simulator (Motion)

L3 Commercial Aviation

September 11, 2018 15

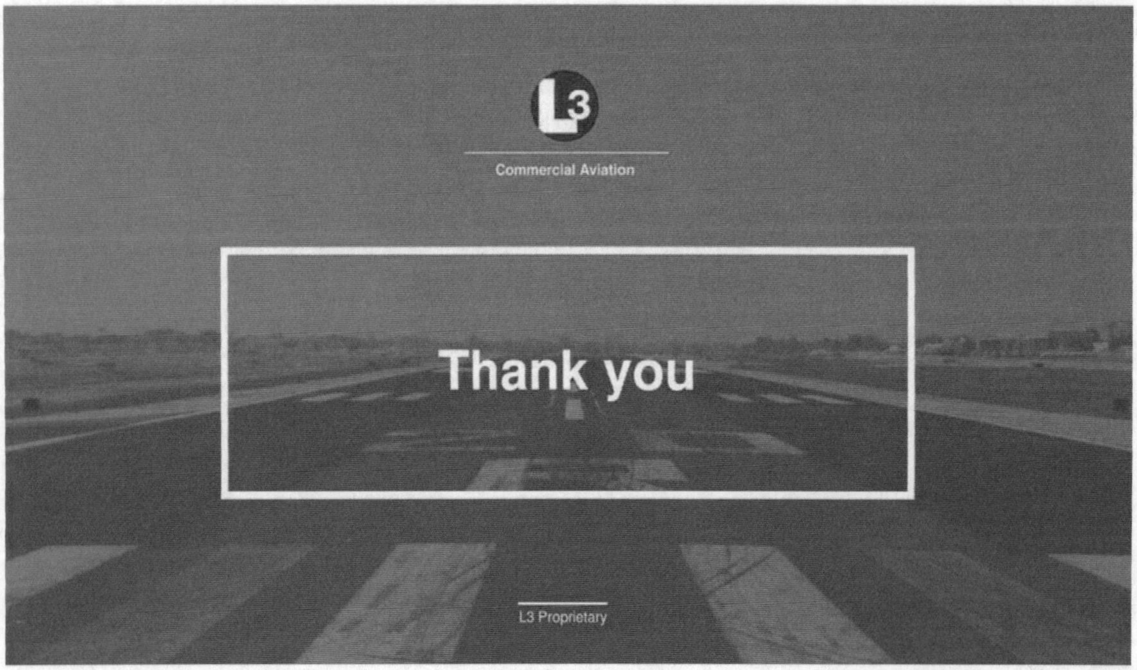

CEFA AVIATION

apats2018 SINGAPORE

HOW COULD WE TURN EACH FLIGHT INTO A TRAINING OPPORTUNITY?

Pierre WANNAZ
Capt. A330/A340
Flight instructor

Singapore - August 29th, 2018

FLIGHT SAFETY FOUNDATION
independent • impartial • international

MARCH 2018 POSITION PAPER

- Levels of **safety** required → meet **growth** demand?
- Maintain sustainable **quantity** & **quality** of pilots?
- Technology + data + experience → maximize **efficiency**?

NO... BUT THERE IS A SOLUTION !

WHAT DO EXPERTS SAY ABOUT DEBRIEFING?

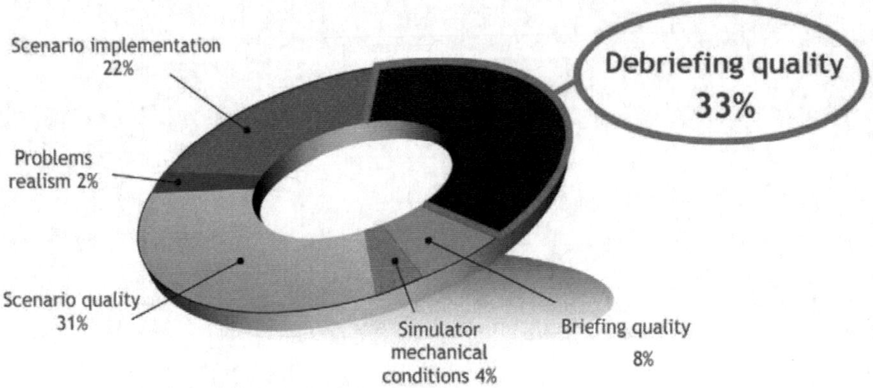

TRAINING EFFICIENCY - FLIGHT SIMULATOR

Source: Swiss HF Guide out of a NASA/AMES study

CAPT. PIERRE WANNAZ AUGUST 29TH 2018

FLIGHT ANALYSIS TODAY

FLIGHT DATA MONITORING (FDM) REPORTS

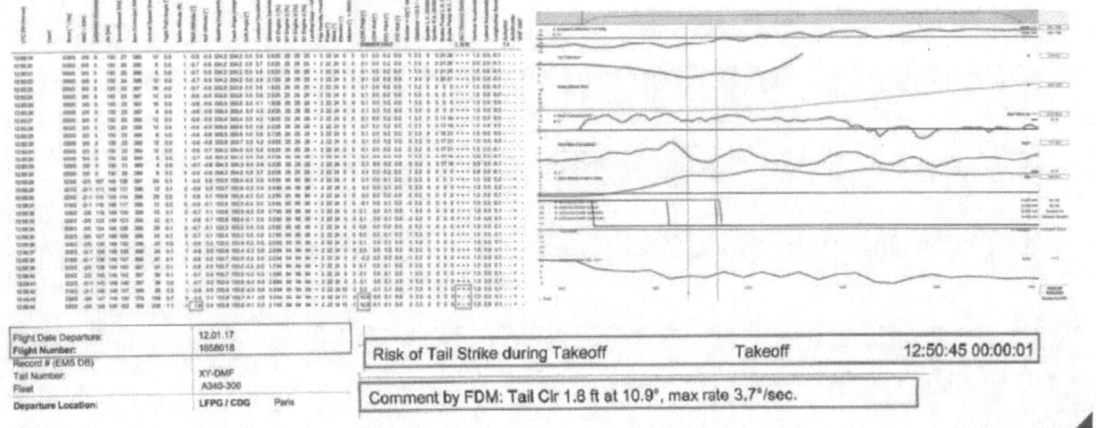

CAPT. PIERRE WANNAZ AUGUST 29TH 2018

CEFA AVIATION

"We do not learn from experience only.
We learn from reflecting on experience..."

self-reflection leads to

20 %

improvement of training performances

CAPT. PIERRE WANNAZ AUGUST 29TH 2018

CEFA AVIATION ➡ **FLIGHT SIMULATORS FOR RECURRENT TRAINING**

3-4 simulator sessions (12-20 hours) a year

BUT

- predefined program (reducing the startle effect)
- not all flight encountered situations are addressed
- expensive (instructor needed/ unproductive day/ travel...)

➔ **Offer a relative LOW training potential improvement at HIGH COST/BENEFIT**

CAPT. PIERRE WANNAZ AUGUST 29TH 2018

CEFA AVIATION

WITH LINE OPERATION

70 (LH) to 850 (SH) flights/year in real life, offering numerous & various situations...

ACTUAL ISSUES/ DIFFICULTIES	global understanding in a dynamic environment debriefings based on pilot memory
ACTUAL POTENTIAL	→ HIGHEST potential for training → HIGHEST potential for safety improvement

at LOW COST/BENEFIT with the right TOOL...

EACH FLIGHT OFFERS A TRAINING OPPORTUNITY

CAPT. PIERRE WANNAZ AUGUST 29TH 2018

apats2018 SINGAPORE

HOW COULD WE TURN EACH FLIGHT INTO A TRAINING OPPORTUNITY?

FLIGHT DEBRIEF - LINE TRAINING

INTRODUCTION VIDEO

Capt. Pierre Wannaz

CEFA AVIATION

A-320

REAL VIDEO ANIMATION

EVENT NOT FLAGGED BY Flight Data Monitoring

COST RELATED AS COOLING IS REQUIRED OR DELAYED FLIGHT

CAPT. PIERRE WANNAZ AUGUST 29TH 2018

CEFA AVIATION

Video Presentation in
Original e-media

CAPT. PIERRE WANNAZ AUGUST 29TH 2018

CEFA AVIATION

THE BENEFITS

FOR PILOTS / CHIEF PILOTS / FLIGHT INSTRUCTORS

1. Confidential access to own flight data only, each pilot being his own gatekeeper

   Animation respecting the anonymity of the involved pilot

2. Flight debriefings based on global/holistic factual evidences

   → lower risk of developing individual techniques
   → EBT brought to individual level

CAPT. PIERRE WANNAZ AUGUST 29TH 2018

CEFA AVIATION

THE BENEFITS

FOR PILOTS / CHIEF PILOTS / FLIGHT INSTRUCTION

3. Better global understanding of occurrences and root causes → eliminate self-doubt → build self-confidence

4. Easier knowledge transfer and sharing of experience between crew members

5. Database of animation with "State of the Art" of difficult Approach

CAPT. PIERRE WANNAZ AUGUST 29TH 2018

CEFA AVIATION

THE BENEFITS

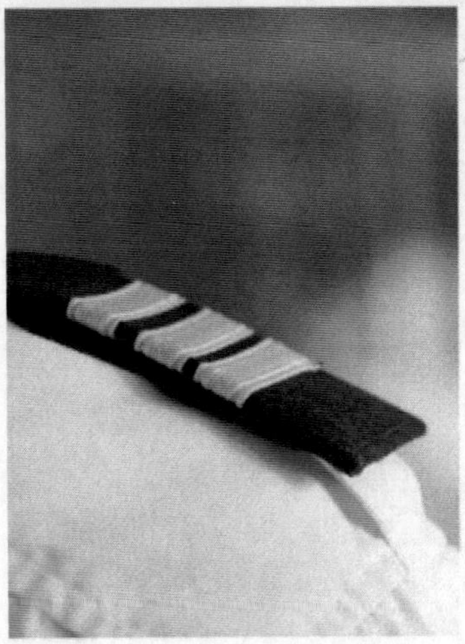

FOR TRAINEES .....A DEBRIEFING SOLUTION

1. Understanding support for dynamic/difficult situations as the crew is still together

2. An analysis tool:

✈ for trainees
→ a solid foundation for a Crew Centered Debriefing

✈ for novice pilots

→ help them to understand own performance using the tools (tablets) that the young generation already use everyday !

CAPT. PIERRE WANNAZ AUGUST 29TH 2018

CEFA AVIATION

THE BENEFITS

FOR SAFETY TEAMS

1. Simple occurrence not flagged as FDM events ignored by the safety system today, can be shown to FDM → animation used as a collaborative tool for flight safety!

JUST CULTURE SUPPORT

REPORTING ENVIRONMENT

'human-in-the-system' safety protection

To create an atmosphere of trust in which people are encouraged, even rewarded, for providing essential safety-related information

CAPT. PIERRE WANNAZ AUGUST 29TH 2018

THE BENEFITS

FOR SAFETY TEAMS

NEW BENEFITS:

2. FDM can provide Big Data statistics for own evaluation vs fleet performance!

3. Trend monitoring of situations before a serious event occurs

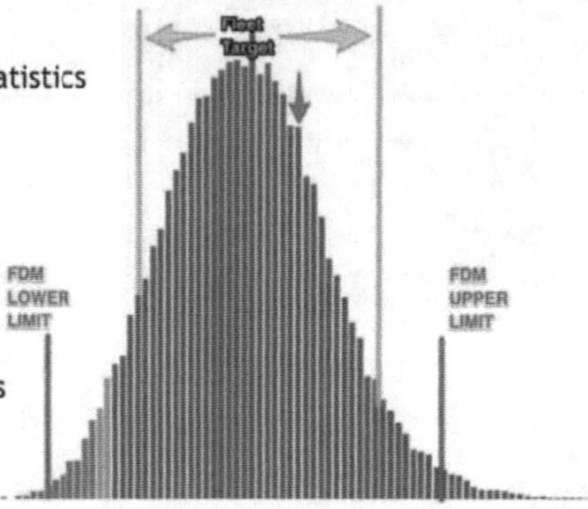

CAPT. PIERRE WANNAZ AUGUST 29TH 2018

RETURN ON EXPERIENCE

"It has revolutionized the company culture regarding the debriefings and the use of flight data.

It has freed pilots' speech!"

Hideo Morioka, Senior Director Safety promotion and Flight Data Analysis

ANA

CAPT. PIERRE WANNAZ AUGUST 29TH 2018

CEFA AVIATION

RESILIENCE

=

CONFIDENCE x COMPETENCE

CAPT. PIERRE WANNAZ AUGUST 29TH 2018

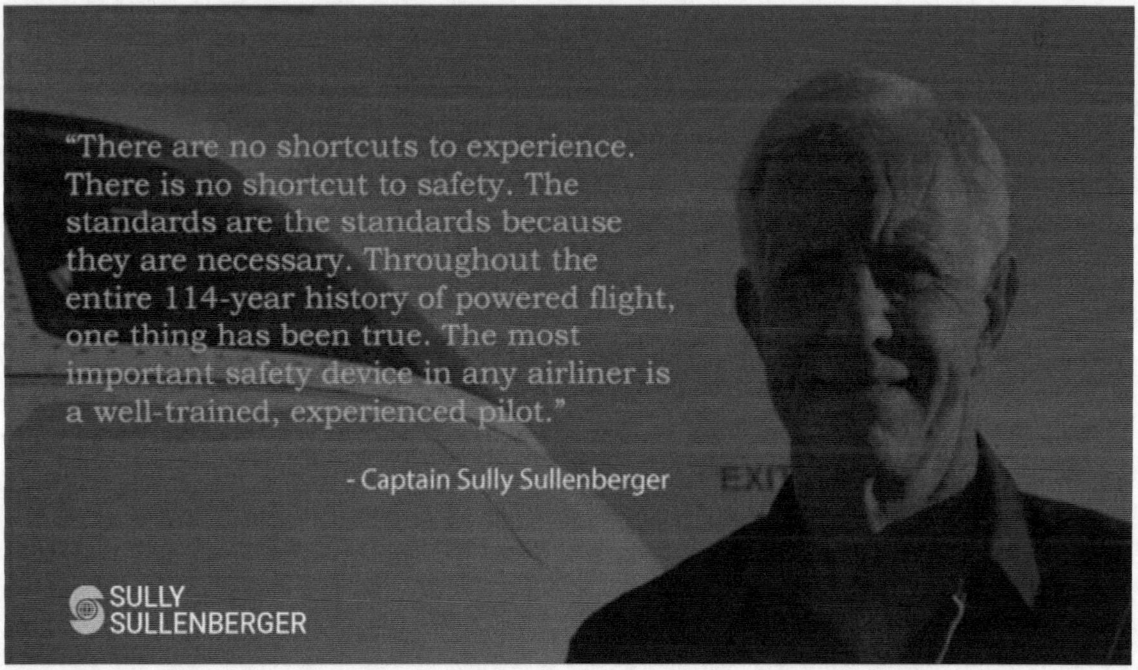

CEFA
AVIATION
apats2018
SINGAPORE
COME & DISCOVER
OUR INNOVATION!
STAND #801

AUTHOR INDEX